Springer Theses

Recognizing Outstanding Ph.D. Research

Aims and Scope

The series "Springer Theses" brings together a selection of the very best Ph.D. theses from around the world and across the physical sciences. Nominated and endorsed by two recognized specialists, each published volume has been selected for its scientific excellence and the high impact of its contents for the pertinent field of research. For greater accessibility to non-specialists, the published versions include an extended introduction, as well as a foreword by the student's supervisor explaining the special relevance of the work for the field. As a whole, the series will provide a valuable resource both for newcomers to the research fields described, and for other scientists seeking detailed background information on special questions. Finally, it provides an accredited documentation of the valuable contributions made by today's younger generation of scientists.

Theses are accepted into the series by invited nomination only and must fulfill all of the following criteria

- They must be written in good English.
- The topic should fall within the confines of Chemistry, Physics, Earth Sciences, Engineering and related interdisciplinary fields such as Materials, Nanoscience, Chemical Engineering, Complex Systems and Biophysics.
- The work reported in the thesis must represent a significant scientific advance.
- If the thesis includes previously published material, permission to reproduce this must be gained from the respective copyright holder.
- They must have been examined and passed during the 12 months prior to nomination.
- Each thesis should include a foreword by the supervisor outlining the significance of its content.
- The theses should have a clearly defined structure including an introduction accessible to scientists not expert in that particular field.

More information about this series at http://www.springer.com/series/8790

Daiki Nishiguchi

Order and Fluctuations in Collective Dynamics of Swimming Bacteria

Experimental Exploration of Active Matter Physics

Doctoral Thesis accepted by
The University of Tokyo, Tokyo, Japan

 Springer

Author
Dr. Daiki Nishiguchi
Department of Physics
Graduate School of Science
The University of Tokyo
Bunkyo-ku, Tokyo, Japan

Supervisor
Prof. Masaki Sano
Tokyo College
Institutes of Advanced Study
The University of Tokyo
Bunkyo-ku, Tokyo, Japan

ISSN 2190-5053   ISSN 2190-5061   (electronic)
Springer Theses
ISBN 978-981-13-9999-2   ISBN 978-981-32-9998-6   (eBook)
https://doi.org/10.1007/978-981-32-9998-6

This Springer imprint is published by the registered company Springer Nature Singapore Pte Ltd.
The registered company address is: 152 Beach Road, #21-01/04 Gateway East, Singapore 189721, Singapore

Supervisor's Foreword

This Ph.D. thesis makes us feel that a new research field called the Experimental Statistical Physics is taking shape after a somewhat lengthy history of the studies of complex systems. When the system consists of many interacting elements, emergent behaviors which cannot be explained by just the sum of elements could often happen. Statistical physics aims to tackle the problems of such emergent behaviors to understand and predict phenomena from theory. The methodology of statistical physics had shown its power for explaining the properties of all kinds of existing materials that consist of molecules, atoms, and electrons. The theory sometimes predicts things which do not exist in nature. On the other hand, experimentalists have designed the predicted systems and created never-existing materials in nature to prove the validity of the theory. Such examples can be found repeatedly in the history of science. Producing ultra-pure silicon crystals and their modification by adding a few impurities were a prominent example of the triumph of science and technology. It dramatically changed their electric properties which lead to modern silicon technology. Theory and experiment have progressed hand in hand, sometimes serendipitous discovery triggered a new theory, and adversely, predictions by theory often motivated experimentalists to verify the theory.

After the success of understanding critical phenomena in the phase transition in the 1970s, statistical physics started to tackle more complex systems such as rhythms, chaos, and turbulence not only in physical systems but also in chemical and biological systems. This movement was somewhat opposite to the direction of going toward more and more microscopic scale to understand the ultimate state of the matter. Nevertheless, communications between theory and experiment have been successful even for attacking complex systems until when they encounter real complex systems such as biological systems and sociological systems. As the composing elements become macroscopic, theory becomes a mere metaphor and experiment often ends in just superficial resemblance. Because each element has more internal degrees of freedom, no one can trust that oversimplification of the elements has any predictive power on the dynamics in biological systems as a whole. Does this mean that scientific efforts going toward macroscopic scales to understand complex systems rather than going to the microscopic scale hit the limit? No, it is not likely that this trend of scientific efforts will stop at the present stage. The breakthroughs seem to be happening here and there. Active matter is one of the breakthroughs in understanding the collective behavior of complex systems, including biological pattern formations. In the 1990s, people in statistical physics started to think that flocking of birds, schooling of fish, and swimming bacterial suspensions may be mathematically modeled by a collection of interacting self-propelled spins. Soon after the proposal of spin-like discrete model, hydrodynamic theory and its predictions have been made. Massive computer simulations have been employed to uncover the richness of collective behaviors depending on symmetry, space dimension, and topology. Although the experimentalists were slower and more modest than theorists, the interaction between theory and experiment has also been developed in the field of active matter. In the 2010s at the latest, a lot of experimental papers on active matter came out. Those include collective motions in artificial microswimmers made of colloids, swimming bacteria in suspension and colony, tissues of eukaryotic cells, biological filaments driven by molecular motors, and so on.

This Ph.D. thesis written by Dr. Daiki Nishiguchi contains one of the state-of-the-art experiments in this field. For example, in the 90s, the hydrodynamic theory predicted the existence of a true long-range ordered state in self-propelled systems in 2D, which is prohibited in the XY spin system at equilibrium. The theory predicted that the giant number fluctuations (abnormal fluctuation abbreviated as GNF) would accompany with that state. However, there had been no experimental evidence of GNF coexisting with long-range ordered states in any systems. Daiki Nishiguchi designed an experiment of swimming bacteria. Here, each bacterium is modified to possess a steric interaction to align with each other. In addition, they can swim only in their heading directions without tumbling, thanks to genetic technology. As a result, the suspension of bacteria exhibited a beautifully aligned long-range order accompanied by GNF for the first time. He has also designed an experiment to test the effect of topological confinement of active matter by using microfabrication technique. The result was yet not as spectacular as topological currents observed in strongly correlated materials. But, it already shows that there are still a lot of possibilities to explore the complex systems if the experiments are designed well. Controlling interactions, propulsion, topological confinement, and spatial dimension combined with various symmetries would create huge variety.

This thesis is to show the beginning of the experimental statistical physics. Fruitful contributions to biology and true understanding of biological systems may follow afterward. But, I believe that this is the necessary step for the science of complex systems. As a whole, this thesis tells us how statistical physics, or unless exaggerated, science proceeds for the understanding of complex systems even including biological systems.

Tokyo, Japan   Prof. Masaki Sano

List of Publications

In reverse chronological order of publication:

1. Daiki Nishiguchi, Igor S. Aranson, Alexey Snezhko, and Andrey Sokolov, "Engineering bacterial vortex lattice via direct laser lithography", Nature Communications, 9, 4486 (2018). DOI: https://doi.org/10.1038/s41467-018-06842-6
2. Daiki Nishiguchi, Junichiro Iwasawa, Hong-Ren Jiang, and Masaki Sano, "Flagellar dynamics of chains of active Janus particles fueled by an AC electric field", New Journal of Physics, 20, 015002 (2018). arXiv: 1709.07756. DOI: https://doi.org/10.1088/1367-2630/aa9b48
3. Daiki Nishiguchi, Ken H. Nagai, Hugues Chaté, and Masaki Sano, "Long-range nematic order and anomalous fluctuations in suspensions of swimming filamentous bacteria", Physical Review E Rapid Communications, 95, 020601(R) (2017). arXiv: 1604.0427. DOI: https://doi.org/10.1103/PhysRevE.95.020601
4. Daiki Nishiguchi and Masaki Sano, "Mesoscopic turbulence and local order in Janus particles self-propelling under an ac electric field", Physical Review E, 92, 052309 (2015). arXiv: 1506.06591. DOI: https://doi.org/10.1103/PhysRevE.92.052309

Papers 1 and 3 include the main contents of this thesis described in Chaps. 5 and 6, respectively. Papers 2 and 4 are introduced in Chap. 4.

Supplementary movies related to this dissertation are available on the web sites of the journals. Especially, the papers 1 and 2 are published in open access journals and are freely available.

The original dissertation, before modification and updates for publication in Springer Theses, is available on UTokyo Repository (DOI: https://www.doi.org/10.15083/00075579).

Acknowledgements

I would like to express my deepest gratitude to my supervisor, Prof. Masaki Sano. He allowed me to freely pursue my ideas and interest. The interdisciplinary atmosphere of his lab enabled me to expand my knowledge and interest through interactions with other members of the lab. His enthusiasm for the search of universal laws in nonequilibrium systems unconsciously but certainly has influenced me. His insightful suggestions from his astonishingly broad perspective always helped me direct my projects toward fantastic outcomes.

I am greatly indebted to Dr. Hugues Chaté for his crucial suggestions and encouragement. My understanding of collective motion was constructed through continuous and stimulating discussion with this leading researcher in active matter physics. I would like to thank Dr. Ken H. Nagai for guiding me into active matter physics and sharing his knowledge and ideas with me. I also thank Prof. Yusuke T. Maeda for exchanging techniques and tips on bacterial experiments with me and for providing transformed bacteria for my experiments. I am thankful to Ms. Haruka Sugiura for teaching me basics of bacterial experiments at the very start-up of my projects.

My special thanks go to Dr. Igor S. Aranson and Dr. Andrey Sokolov. I had a wonderful opportunity to stay at Dr. Aranson's group at Argonne National Laboratory in the U.S. from the beginning of January to the middle of March in 2016. I conducted experiments on bacterial turbulence with them, which greatly expanded my knowledge and understanding of bacterial dynamics. There I learnt a lot from the members of his group not only about physics but also about the attitude toward science.

I would like to express my appreciation to all the past and current members of Sano Laboratory for sharing fruitful time with me and stimulating me. Prof. Kazumasa A. Takeuchi inspired me and kindly took his time to think about possible future directions of my projects with me. Prof. Tetsuya Hiraiwa friendly told me various aspects of active matter. Dr. Ryo Suzuki taught me basics on active matter experiments. Discussion with Dr. Kyogo Kawaguchi was invaluable for me to enhance my understanding of collective motion, biophysics, and statistical physics. I also would like to acknowledge Dr. Yohei Nakayama for his kind experimental and theoretical advice, Dr. Sosuke Ito for broadening my knowledge in nonequilibrium statistical physics, and Dr. Koutarou Otomura for continuous discussion especially on hydrodynamic aspects of microswimmers. Frequent and often casual discussion with Dr. Takaki Yamamoto deepened my thought on various topics in science. I would also like to acknowledge the members of my dissertation committee, Prof. Chikara Furusawa, Prof. Hideo Higuchi, Prof. Nobuyasu Ito, Prof. Hiroshi Noguchi, and Prof. Yutaka Sumino, for refereeing this dissertation and giving me many constructive comments.

Last but not least, I thank my family for supporting and encouraging me not only during my student years but also throughout my life. My beloved wife, Dr. Kana Fuji, has always encouraged me in everything with her wonderful smile. I cannot thank her enough.

Contents

Chapter 1
General Introduction

Abstract A general introduction to the field of active matter physics is presented in this chapter. This thesis is dedicated to experimental studies on collective motion of swimming bacteria from the viewpoint of nonequilibrium statistical physics. Collective motion of self-propelled elements has fascinating properties that are often different from those of orientationally-ordered equilibrium systems due to its intrinsically nonequilibrium nature. To further understand such properties, we explore emergent order and fluctuations in two major classes of collective motion: the Vicsek universality class and active turbulence. Starting from the definitions on self-propulsion and active matter, we briefly review the typical active matter systems and their collective motion in nature and in experiments. By summarizing current situations on the theoretical understanding of collective motion such as state-of-the-art interpretations on giant number fluctuations, we formulate the questions we address in the following chapters. At the end of the chapter, we summarize the organization of this thesis.

Keywords Self-propelled particle · Active matter · Collective motion · Giant number fluctuations · Active turbulence

1.1 Active Matter Physics

Self-propelled objects and their collective motion are so ubiquitous [1]. Flocks of birds fly coherently and sometimes exhibit dynamical deformations as if they were single huge creatures. Schools of fish spontaneously form vortices as we can see in aquariums. Herds of sheep escape from sheepdogs as if they were large 'droplets' with consciousness [2]. On a much larger scale, wildebeest migration spans more than kilometers. When we turn our eyes into the microscopic world, we can find bacteria swarming in their colonies [3]. These fascinating phenomena have captured scientists' imagination.

What is 'self-propulsion'? We need to define this term before moving on to physical arguments. Self-propelled objects are objects whose directions of motion are determined not solely by any external fields but by their internal or hidden degrees

© Springer Nature Singapore Pte Ltd. 2020   1
D. Nishiguchi, *Order and Fluctuations in Collective Dynamics*
of Swimming Bacteria, Springer Theses, https://doi.org/10.1007/978-981-32-9998-6_1

of freedom such as their polarities, their shapes, and their internal chemical signaling.[1] It is also important to note that as a natural requirement of their motility each self-propelled object converts some kind of energy source into kinetic energy.

To better understand collective dynamics of self-propelled elements, many artificial non-living self-propelled particles have been devised, such as nm-scale biofilaments driven by molecular motors [4–6], μm-scale colloids consuming chemical [7, 8], thermal [9], or electric energy [10–13], and mm-scale shaken granular materials [14–16]. These artificial systems are more controllable than living systems and have provided us insight into collective dynamics of motile elements. Thus we now have a wide range of collective phenomena of self-propelled objects, from nanoscales to meter-scale or even kilometer-scales.

From the viewpoint of statistical physics, collective motion of self-propelled elements is a fascinating subject not only because of its ubiquity and possible universality but because it is an intrinsically nonequilibrium phenomenon. Thermodynamics and statistical physics, which deal with macroscopic systems composed of microscopic elements, have had great success in describing passive equilibrium systems. On the other hand, humankind has not yet succeeded in understanding nonequilibrium systems comprehensively. There is indeed some progress in fundamental theories on nonequilibrium states, but they are only applicable in restricted regimes such as linear response regime close to equilibrium, steady states, and fluctuating small systems where stochastic thermodynamics can be applied.

Typical nonequilibrium systems that physicists have ever tried to understand were (a) driven out of equilibrium just by boundary conditions or external fields, or (b) relaxing extremely slowly to equilibrium: Fluid turbulence and convection are driven by shear, pressure gradient, or temperature gradient set at boundaries; electrophoresis and electric conduction are triggered by an electric field applied at boundaries; glass and granular systems have enormously long relaxation time that cannot be achieved in realistic time scales. In these systems, energy is injected at boundaries or just as an initial condition, and then it eventually dissipates in bulk. On the contrary, in the case of collective motion of self-propelled elements, both energy injection and dissipation take place in bulk. Each element transforms some sort of energy into kinetic energy, which means energy injection occurs at the smallest scale of the system. This injected energy is then transferred toward larger spatial scales through interactions among self-propelled objects and eventually dissipate again in bulk, leading to the emergence of spectacular dissipative structures. As such, collective motion of self-propelled elements is one of the most difficult but interesting problems in nonequilibrium statistical physics.

The term 'active matter' is frequently used for denoting such a field of study, but its definition is rather obscure compared with self-propulsion. This word is often

---

[1]One might oppose to this definition by raising chemotaxis and collective motion. In these cases, the directions of motion are indeed 'modified' by the external chemical gradient or by their neighbors. However, each self-propelled object can still move at a certain direction even without such external fields, and thus these are upper-level phenomena exhibited by self-propelled objects in response to the external fields.

used for describing a group of self-propelled elements, although some people call a single self-propelled object, a single living organism, and even a single cell by active matter. Hence 'active matter' can be recognized as a hypernym of self-propelled objects. Anyhow, there is no doubt that endeavor to understand collective motion of self-propelled objects is at the heart of 'active matter physics'.

At the end of this section, we would like to mention the role of active matter physics in relation to other disciplines in biological/bio-related science. What is life? What are principles behind biological activities? Answering these crucial questions is one of the ultimate goals in science. Biological activities range in vast length scales, and scientists have been involved in these phenomena from microscale to macroscale: gene expression, molecular motors, cytoplasmic streaming, cell motility, development, morphogenesis, ecosystem, etc. Not only biologists but also physicists have been working on these questions. Along this line, active matter physics can be recognized as a new discipline tackling mesoscale biological/biomimetic activities to connect knowledge obtained in other fields.

1.2 Exploration of Universality in Collective Motion

When we start to think about collective motion from the viewpoint of statistical physics, simple but crucial question comes to us: How and why can birds flock together, even though each bird cannot see the whole flock from inside? Of course, the same question can be raised for other systems such as wildebeests, fish and bacteria. In other words, how can such coherent collective motion of self-propelled elements emerge from their local short-ranged interactions? This reminds us of basic mathematical models in equilibrium statistical physics such as the Ising model and the XY model, and of their great capabilities of capturing universality.

With the success of mathematical models in equilibrium systems in mind, in 1995, Tamás Vicsek et al. devised a flying XY model—now called the Vicsek model— in which each spin moves to its own spin direction [17]. They demonstrated a transition from a disordered random state to a highly ordered state even in two dimensions (2D). Only four months after Vicsek's paper came out, John Toner and Yuhai Tu published their seminal paper [18] proving that this ordered state have true long-range order which cannot be achieved in equilibrium 2D systems as stated in the Mermin-Wagner theorem [19]. As such, this transition is a strikingly novel type of nonequilibrium phase transition that physicists have never thought about. It has attracted much attention from physicists and motivated them to search for possibly universal properties of collective motion.

In these two decades after the Vicsek model was introduced, following the seminal works by Toner, Tu, Ramaswamy et al. [14, 18, 20–22], evidence for such universality has been indeed provided by many theoretical studies and large scale numerical simulations on simple flocking models where orientationally-ordered flocking states emerge as a result of competition between local alignment and noise [23–28]. It is now understood that the nonequilibrium phase transition from incoherent random motion to orientationally-ordered collective motion in such models is described by a phase-separation between a disordered 'gas' state and an orientationally-ordered 'liquid' state separated by a coexistence phase whose nature depends on the symmetries of the system [29, 30]. This homogeneous but highly fluctuating liquid phase observed in such models, which we call the Toner-Tu-Ramaswamy (TTR) phase hereafter, has abundant new physics. It is characterized by unique properties often distinctively different from those of equilibrium orientationally-ordered phases as seen in the classical XY model and nematic liquid crystals. As we have already mentioned above, under a certain condition it can develop true long-range order even in 2D, which is prohibited in equilibrium systems by the Mermin-Wagner theorem [19]. Among all the theoretical predictions, the most striking one is anomalously-large density fluctuations, or 'giant number fluctuations' (GNF), generated from both the algebraic correlations in this spontaneously symmetry broken phase and the crucial coupling between the orientational order and the density field [14, 18, 20–22]. Because GNF represent fundamental mathematical properties inherent in the TTR phase, GNF are now considered and treated as a hallmark of collective motion. These simple flocking models exhibiting the TTR phase are now considered to constitute 'the Vicsek universality class'.

Although series of theoretical and numerical works have clarified the properties of the TTR phase and the Vicsek universality class, no experiments have demonstrated such universality in a fully convincing way due to experimental difficulties and many pitfalls. Due to relatively easy experimental accessibility of number fluctuations, many experiments have reported 'GNF', although they are not their main claims. Several experimental systems were elaborated and employed for detecting 'GNF', such as biofilaments driven by molecular motors [6], colloids consuming electric energy [12], shaken granular materials [15, 16, 31], monolayers of fibroblast cells [32], and common bacteria [3, 33]. However, none of these experiments could exhibit the TTR phase with GNF, as we will explain in detail in Sect. 2.6. The 'GNF' reported there were measured *out of the TTR phase* where necessary conditions for discussing GNF in the sense of the TTR phase were overlooked, and were originated from uninteresting clustering or boundary effects. Hence some physicists still believe that GNF as predicted by Toner, Tu, Ramaswamy et al. can trivially be obtained in active matter systems, which is a widespread troublesome misconception on GNF.

In actual experimental systems on microswimmers such as usual bacteria and self-propelled colloids, what we observe instead of the TTR phase are spatio-temporally chaotic phases with short-range orientational order, which are now termed 'active turbulence' or 'mesoscopic turbulence' in analogy with classical fluid turbulence [33–38]. This another class of collective motion stems from hydrodynamic flow created by these microswimmers that destabilizes TTR-like globally ordered states or aligned configuration of microswimmers, whose effect was *not* present in the simple flocking models [39–43]. Thanks to experimental accessibility of bacterial turbulence, or active turbulence in bacterial suspensions, many experimental works and corresponding theoretical modeling were performed, and active turbulence is now well described by hydrodynamic equations [33, 38, 44]. Although active turbulence cannot exhibit global order as seen in the Vicsek universality class, recent experiments have demonstrated that we can still obtain a kind of 'order' in active turbulence as a result of the interplay between active turbulence and boundaries: directed transport of a wedge [45], directed rotation of microscopic gears [46], spontaneous spiral vortex formation under circular confinement [47, 48], ferromagnetic and antiferromagnetic vortex lattice formation [49], directed collective motion under channel confinement [50], etc. How can we obtain such 'order' out of fluctuating chaotic regime? The study on active turbulence is evolving into the next stage.

In this thesis, by using suspensions of swimming bacteria as model active matter systems, we present the experimental studies on two major classes of collective motion mentioned above: the homogeneous but highly fluctuating orientationally-ordered phase (the TTR phase) and active turbulence. Our experimental system with swimming filamentous bacteria in quasi-two-dimensions falls into the Vicsek universality class, which is the first unambiguous experimental realization of the TTR phase [51]. This observation gives us deep insight into what is necessary for the emergence of the TTR phase. The other experiment on active turbulence treats the interplay between bacterial turbulence and periodic obstacles [52]. We obtain a resonant state at certain periodicity, which gives us an explicit answer to the question on what happens when bulk active turbulence meets obstacles. The obtained results highlight the existence of characteristic length scale in bacterial turbulence and its importance for the emergence of order out of chaos. Through these studies, we address the state-of-the-art understanding of the emergent order and the universality in active matter systems.

1.3 Organization of the Thesis

The following of this thesis is divided into two parts. The first part, Chaps. 2 and 3, focuses on the Toner-Tu-Ramaswamy phases of collective motion of self-propelled elements. The second part, Chaps. 4 and 5, describes the dynamics of bacterial turbulence.

In Chap. 2, we review existing theoretical and numerical studies on collective motion of self-propelled elements. Specifically, we introduce Vicsek-style models and corresponding hydrodynamic theories. The outcomes of these numerical and theoretical studied suggest universality in collective motion. At the end of this chapter, we summarize experimental studies that have strived to find such universality, and point out the crucial discrepancies between those experimental works and the theoretical/numerical studies.

Being aware of such problems, we present our experimental study on collective motion of filamentous bacteria in Chap. 3. Our experimental results clearly demonstrate that this system falls into the Vicsek universality class. This gives the first experimental realization of the Toner-Tu-Ramaswamy phases, which have been theoretically predicted and studied in depth but have never been observed experimentally due to many pitfalls. Our findings give insights on what is necessary for the emergence of the Toner-Tu-Ramaswamy phases in real experimental systems, and provide clues for future theoretical/experimental development.

From Chap. 4, we move on to investigate active turbulence, in particular, bacterial turbulence. In Chap. 4, we review basic properties of bacterial turbulence and existing experimental works. Although bacterial turbulence exhibits chaotic behavior, ordered coherent motion was found under confinement. However, such confined experiments could not be run for a long time, and it remained elusive what happens when unconstrained bacterial turbulence encounters some structures.

In Chap. 5, we present our experimental study on self-organization of bacterial turbulence in the presence of periodic structures. We explore how order emerges as a result of the interplay between them by devising microarrays of pillar lattices in which we can simultaneously measure behavior of bacterial turbulence both in the bulk and in the periodic structures with various periodicities. At the appropriate periodicity, bacterial turbulence self-organize in highly stabilized antiferromagnetic vortex lattices. The obtained results highlight the existence of characteristic length scale and the importance of such a length scale and boundary conditions for the emergence of order out of fluctuating chaotic bacterial turbulence.

Finally, in Chap. 6, we aim to clarify the significance and implications of our studies. We conclude by mentioning possible future directions of active matter physics.

References

1. Vicsek T, Zafeiris A (2012) Collective motion. Phys Rep 517:71–140
2. Ginelli F, Peruani F, Pillot M-H, Chaté H, Theraulaz G, Bon R (2015) Intermittent collective dynamics emerge from conflicting imperatives in sheep herds. Proc Natl Acad Sci USA 112(41):12729–12734
3. Zhang HP, Be'er A, Florin E-L, Swinney HL (2010) Collective motion and density fluctuations in bacterial colonies. Proc Natl Acad Sci USA 107(31):13626–13630
4. Schaller V, Weber C, Semmrich C, Frey E, Bausch AR (2010) Polar patterns of driven filaments. Nature 467(7311):73–77
5. Sumino Y, Nagai KH, Shitaka Y, Tanaka D, Yoshikawa K, Chaté H, Oiwa K (2012) Large-scale vortex lattice emerging from collectively moving microtubules. Nature 483:448–452
6. Schaller V, Bausch AR (2013) Topological defects and density fluctuations in collectively moving systems. Proc Natl Acad Sci USA 110(12):4488–4493
7. Palacci J, Cottin-Bizonne C, Ybert C, Bocquet L (2010) Sedimentation and Effective Temperature of Active Colloidal Suspensions. Phys Rev Lett 105(8):088304
8. Ginot F, Theurkauff I, Levis D, Ybert C, Bocquet L, Berthier L, Cottin-Bizonne C (2015) Nonequilibrium equation of state in suspensions of active colloids. Phys Rev X 5(1):011004
9. Jiang H-R, Yoshinaga N, Sano M (2010) Active motion of a janus particle by self-thermophoresis in a defocused laser beam. Phys Rev Lett 105(26):268302
10. Suzuki R, Jiang H-R, Sano M (2011) Validity of fluctuation theorem on self-propelling particles. arXiv:1104.5607
11. Nishiguchi D, Sano M (2015) Mesoscopic turbulence and local order in Janus particles self-propelling under an ac electric field. Phys Rev E 92(5):052309
12. Bricard A, Caussin J-B, Desreumaux N, Dauchot O, Bartolo D (2013) Emergence of macroscopic directed motion in populations of motile colloids. Nature 503(7474):95–98
13. Nishiguchi D, Iwasawa J, Jiang H-R, Sano M (2018) Flagellar dynamics of chains of active Janus particles fueled by an AC electric field. New J Phys 20:015002

14. Ramaswamy S, Simha RA, Toner J (2003) Active nematics on a substrate: giant number fluctuations and long-time tails. Eur Lett 62(2):196–202
15. Deseigne J, Dauchot O, Chaté H (2010) Collective motion of vibrated polar disks. Phys Rev Lett 105(9):098001
16. Kumar N, Soni H, Ramaswamy S, Sood AK (2014) Flocking at a distance in active granular matter. Nat Commun 5:4688
17. Vicsek T, Czirók A, Ben-Jacob E, Cohen I, Shochet O (1995) Novel type of phase transition in a system of self-driven particles. Phys Rev Lett 75(6):1226
18. Toner J, Tu Y (1995) Long-range order in a two-dimensional dynamical XY model: how birds fly together. Phys Rev Lett 75(23):4326–4329
19. Mermin ND, Wagner H (1966) Absence of ferromagnetism or antiferromagnetism in one- or -two-dimensional isotropic Heisenberg models. Phys Rev Lett 17(22):1133
20. Toner J, Tu Y (1998) Flocks, herds, and schools: a quantitative theory of flocking. Phys Rev E 58(4):4828–4858
21. Toner J (2012) Reanalysis of the hydrodynamic theory of fluid, polar-ordered flocks. Phys Rev E 86(3):031918
22. Toner J, Tu Y, Ramaswamy S (2005) Hydrodynamics and phases of flocks. Ann Phys 318:170–244
23. Marchetti MC, Joanny JF, Ramaswamy S, Liverpool TB, Prost J, Rao M, Simha RA (2013) Hydrodynamics of soft active matter. Rev Mod Phys 85(3):1143–1189
24. Grégoire G, Chaté H (2004) Onset of collective and cohesive motion. Phys Rev Lett 92(2):025702
25. Chaté H, Ginelli F, Grégoire G, Peruani F, Raynaud F (2008) Modeling collective motion: Variations on the Vicsek model. Eur Phys J B 64:451–456
26. Chaté H, Ginelli F, Montagne R (2006) Simple model for active nematics: Quasi-long-range order and giant fluctuations. Phys Rev Lett 96(18):180602
27. Ginelli F, Peruani F, Bär M, Chaté H (2010) Large-scale collective properties of self-propelled rods. Phys Rev Lett 104(18):184502
28. Ngo S, Peshkov A, Aranson IS, Bertin E, Ginelli F, Chaté H (2014) Large-scale chaos and fluctuations in active nematics. Phys Rev Lett 113(3):038302
29. Solon AP, Tailleur J (2013) Revisiting the flocking transition using active Spins. Phys Rev Lett 111(7):078101
30. Solon AP, Chaté H, Tailleur J (2015) From phase to microphase separation in flocking models: the essential role of nonequilibrium fluctuations. Phys Rev Lett 114(6):068101
31. Narayan V, Ramaswamy S, Menon N (2007) Long-lived giant number fluctuations in a swarming granular nematic. Science 317(July):105(New York, N.Y.)
32. Duclos G, Garcia S, Yevick HG, Silberzan P (2014) Perfect nematic order in confined monolayers of spindle-shaped cells. Soft Matter 10(14):2346–2353
33. Wensink HH, Dunkel J, Heidenreich S, Drescher K, Goldstein RE, Löwen H, Yeomans JM (2012) Meso-scale turbulence in living fluids. Proc Natl Acad Sci USA 109(36):14308–14313
34. Dombrowski C, Cisneros L, Chatkaew S, Goldstein RE, Kessler JO (2004) Self-concentration and large-scale coherence in bacterial dynamics. Phys Rev Lett 93(9):098103
35. Cisneros LH, Cortez R, Dombrowski C, Goldstein RE, Kessler JO (2007) Fluid dynamics of self-propelled microorganisms, from individuals to concentrated populations. Exp Fluids 43(5):737–753
36. Sokolov A, Aranson IS, Kessler JO, Goldstein R (2007) Concentration Dependence of the Collective Dynamics of Swimming Bacteria. Phys Rev Lett 98(15):158102
37. Sokolov A, Aranson IS (2012) Physical properties of collective motion in suspensions of bacteria. Phys Rev Lett 109(24):248109
38. Dunkel J, Heidenreich S, Drescher K, Wensink HH, Bär M, Goldstein RE (2013) Fluid dynamics of bacterial turbulence. Phys Rev Lett 110(22):228102
39. Subramanian G, Koch DL (2009) Critical bacterial concentration for the onset of collective swimming. J Fluid Mech 632:359

40. Ishikawa T, Sekiya G, Imai Y, Yamaguchi T (2007) Hydrodynamic interactions between two swimming bacteria. Biophys J 93(6):2217–2225
41. Saintillan D, Shelley MJ (2007) Orientational order and instabilities in suspensions of self-locomoting rods. Phys Rev Lett 99(5):058102
42. Saintillan D, Shelley MJ (2008) Instabilities and pattern formation in active particle suspensions: Kinetic theory and continuum simulations. Phys Rev Lett 100(17):178103
43. Saintillan D, Shelley MJ (2012) Emergence of coherent structures and large-scale flows in motile suspensions. J R Soc Interface 9(68):571–85
44. Dunkel J, Heidenreich S, Bär M, Goldstein RE (2013) Minimal continuum theories of structure formation in dense active fluids. New J Phys 15:045016
45. Kaiser A, Peshkov A, Sokolov A, ten Hagen B, Löwen H, Aranson IS (2014) Transport powered by bacterial turbulence. Phys Rev Lett 112(15):158101
46. Sokolov A, Apodaca MM, Grzybowski BA, Aranson IS (2010) Swimming bacteria power microscopic gears. Proc Natl Acad Sci USA 107(3):969–974
47. Wioland H, Woodhouse FG, Dunkel J, Kessler JO, Goldstein RE (2013) Confinement stabilizes a bacterial suspension into a spiral vortex. Phys Rev Lett 110(26):268102
48. Lushi E, Wioland H, Goldstein RE (2014) Fluid flows created by swimming bacteria drive self-organization in confined suspensions. Proc Natl Acad Sci USA 111(27):9733–9738
49. Wioland H, Woodhouse FG, Dunkel J, Goldstein RE (2016) Ferromagnetic and antiferromagnetic order in bacterial vortex lattices. Nat Phys 12:341–345
50. Wioland H, Lushi E, Goldstein RE (2016) Directed collective motion of bacteria under channel confinement. New J Phys 18(7):075002
51. Nishiguchi D, Nagai KH, Chaté H, Sano M (2017) Long-range nematic order and anomalous fluctuations in suspensions of swimming filamentous bacteria. Phys Rev E 95(2):020601(R)
52. Nishiguchi D, Aranson IS, Snezhko A, Sokolov A (2018) Engineering bacterial vortex lattice via direct laser lithography. Nat Commun(9)4486:1–8

Chapter 2
Standard Models on Collective Motion

Abstract We start by reviewing previous theoretical and numerical studies on collective motion of self-propelled elements. Especially, we comprehensively review standard models on collective motion: Vicsek-style models and corresponding hydrodynamic theories. Then existing experimental works are carefully examined based on the theoretical understandings. We stress that giant number fluctuations rooted in the spontaneous rotational symmetry breaking need to be discussed in the homogeneous long-range ordered phase without any clusters. We conclude this chapter by mentioning the gaps between those previous experiments and the theoretical predictions. We clarify what is required to observe experimentally to test the theoretical predictions on the Vicsek universality class.

Keywords Collective motion · Vicsek model · Active nematics
Self-propelled rods · Toner-Tu-Ramaswamy phase

2.1 Overview of Standard Models

If generic and robust features of active matter systems exist, they should also be present in the emergent phenomena observed in simple models. We expect, from the statistical physics point of view, that coarse-grained simple models possess universality regardless of details of actual flocks such as interaction rules. In this spirit, in 1995, Tamás Vicsek et al. devised a flying XY model, which is now called "the Vicsek model". Later theoretical studies and large-scale numerical simulations on this simple model have revealed its fundamental properties, most of which are shared by its variant models with different symmetries. Hence the Vicsek model is now regarded as the most basic model on collective motion.

In the following of this chapter, we introduce three basic classes of models of collective motion: the Vicsek model, active nematics, and self-propelled rods. These model can be classified by symmetries of motility and interaction of their self-propelled elements as shown in Table 2.1. We also introduce corresponding hydrodynamic theories. These continuum descriptions are analytically tractable and give us clear and intuitive understandings of collective motion.

© Springer Nature Singapore Pte Ltd. 2020

D. Nishiguchi, *Order and Fluctuations in Collective Dynamics of Swimming Bacteria*, Springer Theses, https://doi.org/10.1007/978-981-32-9998-6_2

Table 2.1 Classification of models of collective motion

	The Vicsek model	Active nematics	Self-propelled rods
Motility	Polar	Apolar	Polar
Interaction	Ferromagnetic	Nematic	Nematic

All of the models and the hydrodynamic descriptions exhibit the same, or at least similar, phenomenology and we can regard that they constitute a kind of nonequilibrium universality class, 'the Vicsek universality class'.

2.2 The Original Vicsek Model

2.2.1 Definition

The Vicsek model is a discrete-time stochastic model on point-like overdamped self-propelled particles with constant speed v_0 and with short-ranged interaction. Its dynamics in two spatial dimensions is given as,[1]

Vicsek model

$$\theta_j^{t+1} = \arg \sum_{k \sim j} e^{i\theta_k^t} + \eta_j^t, \tag{2.1}$$

$$r_j^{t+1} = r_j^t + v_0 e_{\theta_j^{t+1}}. \tag{2.2}$$

Here the single time step of the dynamics is normalized as 1, r_j^t and θ_j^t represent the position and the direction of the j-th particle at time t respectively, $e_{\theta_j^{t+1}}$ is a unit vector directing the θ_j^{t+1}-direction, arg is a function returning the argument of a complex number, η_j^t is a white noise uniformly distributed on $[-\eta/2, +\eta/2]$ with the strength $\eta (> 0)$, and $\sum_{k \sim j}$ indicates a summation over the particles within the interaction radius R from the j-th particle including itself. Because it is easy to generalize its dynamics to higher spatial dimensions and experimentally two dimensional (2D) models are more realistic, we hereafter restrict ourselves to treat two-dimensional models in an $L \times L$ plane with periodic boundary conditions otherwise explicitly stated.

---

[1]In the definition here, θ_j^t is updated prior to the update of r_j^t. It is known that the Vicsek model is robust against the order of updating these two variables.

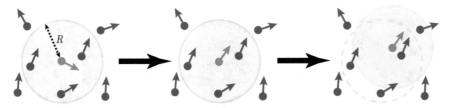

Fig. 2.1 Schematic figure of the dynamics of the Vicsek model. Each particle tries to reorient itself and moves to the average direction of their neighbors closer than a certain interaction radius R. Although in this figure only the red particle is evolving in time, in the actual algorithm all particles are updated simultaneously

The schematic picture of this dynamics is depicted in Fig. 2.1. At each time step, each particle looks around its neighbors, and then aligns and moves to the average direction. However, there exist noise because each particle cannot exactly calculate the average direction or some intrinsic/extrinsic fluctuations are present. Therefore, in this model, alignment competes with noise just like ferromagnetic interaction competes with thermal fluctuations in the Ising model.

From the next subsection, we will look at abundant properties of the Vicsek model.

2.2.2 Physical Properties

Order-disorder phase transition

Which parameters play fundamental roles when we think about the behavior of the Vicsek model? The Vicsek model has following parameters: the self-propulsion speed v_0, the interaction radius R, the noise strength η, and the mean number density of particles $\rho_0 := N/L^2$, where N is the number of particles and L is the linear size of the system. In the case of $v_0 > R$, some particles pass by each other without any interaction, which is unrealistic and unphysical. Hence we have to take $v_0 \leq R$, and it has been verified that under this condition the choice of v_0 and R does not affect macroscopic results. So we can assume, for example, $R = 1$ and $v_0 \simeq 1/2$ without loss of generality. Therefore, the remaining significant parameters in the Vicsek model are η and ρ_0. The noise strength η and the number density ρ in the Vicsek model correspond to temperature and strength of the interaction in equilibrium models such as the Ising model and the classical XY model.

As a result of the competition between alignment and noise, the Vicsek model exhibits an order-disorder phase transition as depicted in Figs. 2.2 and 2.3. At sufficiently high noise strength or at low number density of particles (i.e. weak interactions), the system of self-propelled particles is disordered as a whole with no preferred directions. As we decrease the noise strength or increase the number density, it transits to an ordered flocking state in which the whole system migrates toward a certain direction. In other words, continuous rotational symmetry gets broken in

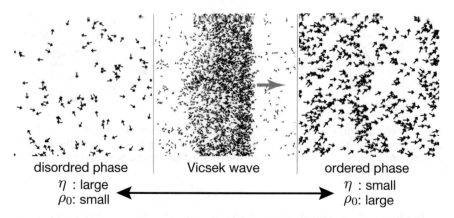

disordred phase | Vicsek wave | ordered phase
η : large | ⟵⟶ | η : small
ρ_0: small | | ρ_0: large

Fig. 2.2 Phases of the Vicsek model. As we decrease the noise strength η or increase the number density of particles ρ_0, we observe a phase transition from the disordered phase to the ordered phase. At the transitional region, coexistence of the ordered region and the disordered region is observed, which indicates the first order nature of this phase transition. In this coexistence phase, band structures are formed orthogonal to the average direction of the particles in the ordered region. Hence these bands propagate and they are called 'Vicsek wave'. In each phase, only some fractions of particles are shown for clarity. Figures of each phase are kindly provided by Dr. Hugues Chaté

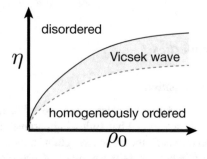

Fig. 2.3 Schematic figure of a phase diagram of the Vicsek model. Depending on the values of the noise strength η and the mean number density ρ_0, the Vicsek model takes the disordered phase, the Vicsek wave, or the ordered phase. The phase boundary at the onset of the order (black solid line) is almost $\eta \sim \sqrt{\rho_0}$. From the arguments at the hydrodynamic description level, the black solid line and the red dashed line represent the linear stability lines where the ordered state and the disordered state gets unstable respectively

this phase transition. This ordered phase, after waiting for a sufficiently long time before it reaches steady states, is spatially homogeneous without any clusters.[2]

At the transitional region, there exists a coexistence phase. The ordered region appears in the disordered phase as band structures. These bands span orthogonally to the average direction of the particles inside the bands, and therefore the bands propagate in the system. These banding structures are called 'Vicsek wave'.

---

[2]Note that, if you do not wait until the system relaxes to steady states, there exist apparent clusters.

The phase boundary at the onset of the order (the black solid line in Fig. 2.3) can be approximated by estimating the mean free path l_f and the persistence length l_p of self-propelled particles. In d-dimensional systems, the mean free path, or the mean interparticle distance is $l_f \sim 1/\rho_0^{1/d}$. The persistence length of each particle should be inversely proportional to the noise strength η, thus we obtain $l_p \sim 1/\eta$. At the transition point, these two length scales are expected to become comparable and thus from $l_f \sim l_p$ we obtain,

$$\eta \sim \rho_0^{1/d}. \tag{2.3}$$

Therefore, when $d = 2$, the phase boundary between the disordered and the ordered phases scales as $\eta \sim \sqrt{\rho_0}$. This rough estimation was confirmed to be valid by the numerical simulations [1].

The phase transition can be quantified by the polar order parameter,

$$\varphi^t = \frac{1}{N} \left| \sum_{j=1}^{N} e^{i\theta_j^t} \right| \tag{2.4}$$

which is defined in the same way as the magnetization in the classical XY model. This φ^t represents the average direction of motion of the system. Because φ^t fluctuates in time, the temporally averaged value $\langle \varphi \rangle_t$ in steady states can capture how ordered the system is. It is important to note that, although $\langle \varphi \rangle_t$ in the disordered phase is 0 in a large system size limit, in a finite system size with N particles $\langle \varphi \rangle_t$ has positive values even in the disordered phase and it scales as $\langle \varphi \rangle_t \propto 1/\sqrt{N}$ due to the law of large numbers.

Historically speaking, the Vicsek wave phase was not discovered for a while because the limited system size of earlier simulations had obscured the transitional region. When Vicsek introduced his model in 1995, he regarded the phase transition as being second-order because the accessible system size was small [2]. However, as depicted in Fig. 2.4, later large-scale simulations with finite-size scaling and with analysis using the Binder cumulant have clearly demonstrated that the transition is first-order (discontinuous) [1, 3].[3]

Giant number fluctuations (GNF)

Giant number fluctuations (GNF) are now regarded as a landmark property of the collective motion that belongs to the Vicsek universality class, because it reflects fundamental mathematical properties of the ordered phase: breaking of rotational symmetry and coupling between orientation and density. GNF were first predicted in seminal works on hydrodynamic theories by John Toner, Yuhai Tu, Sriram Ramaswamy et al., which we will detail later [4–8].

---

[3]In the first Vicsek's paper [2], the number of simulated particles was 10^4 at the largest. On the other hand, in [1], the number of particles was more than 5×10^5.

Fig. 2.4 Schematic figure of
the behavior of $\langle\varphi\rangle_t$ as a
function of noise intensity η
at the transition point
become discontinuous as we
increase the system size.
Such discontinuity has been
quantified by using the
Binder cumulant [1]

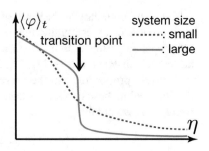

The definition of the number fluctuations is the following. Let us consider the number of the observed particles $N(t)$ inside a certain observation box at time t. This $N(t)$ is a stochastic variable. From the time series of $N(t)$, we can calculate its mean value $\langle N\rangle_t$ and its standard deviation $\Delta N = \sqrt{\langle(N - \langle N\rangle_t)^2\rangle_t}$, where $\langle\,\rangle_t$ denotes a temporal average over sufficiently long period.[4] This standard deviation ΔN is called 'number fluctuations'. In steady states of ergodic systems, which we usually discuss, temporal averages taken above can be replaced by spatial average. However, temporal averages are often experimentally more accessible than spatial averages due to experimental constraints such as inhomogeneity of the setup, limited observation field of view, etc. Hence here we introduce number fluctuations by using temporal averages, and hereafter we omit the subscript t.

Both $\langle N\rangle_t$ and ΔN increase with increasing the observation box size as depicted in Fig. 2.5. In a large box size limit, ΔN scales asymptotically as $\Delta N \propto \langle N\rangle^\alpha$. In equilibrium systems and random systems, the central limit theorem assures that the fluctuations are 'normal' with the exponent $\alpha = 0.5$. On the other hand, the homogeneous ordered collective phase of the Vicsek model exhibits 'giant number fluctuations (GNF)' with the exponent $\alpha > 0.5$. Large scale simulations indicate that the exponent is close to 0.8 for the Vicsek model, whose value is also derived analytically from the hydrodynamic theory by Toner and Tu under some approximations [4–7, 9]. Of course, in the disordered phase of the Vicsek model, we obtain normal fluctuations with $\alpha = 0.5$.

Physical and intuitive interpretation of GNF

From the physical and mathematical point of view, GNF in the ordered phase of the Vicsek model are a manifestation of spontaneous symmetry breaking of rotational symmetry in the collective state. In the transition from the disordered phase to the ordered phase, continuous rotational symmetry of the system gets spontaneously broken. Such spontaneous breaking of continuous symmetry is accompanied by long-wavelength fluctuations, which is called 'the Nambu-Goldstone mode (NG mode)'. In the ordered phase with broken rotational symmetry, each particle can be simultaneously rotated by the same angles *without any frustrations nor any energy cost*, which means there is no restoring force.

---

[4]The period T should be long enough so that the $N(t)$ and $N(t + T)$ are decorrelated.

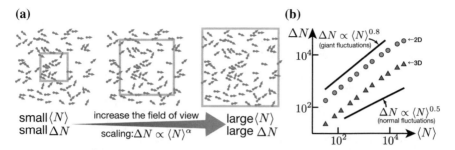

Fig. 2.5 **a** Schematic image of how to calculate number fluctuations. Both number fluctuations ΔN and mean values $\langle N \rangle$ increase as increasing the field of view. There is an asymptotic scaling relation $\Delta N \propto \langle N \rangle^{\alpha}$. **b** The exponent α can be extracted from the log-log plot. In equilibrium systems and random systems, normal fluctuations with $\alpha = 0.5$ are obtained due to the central limit theorem. On the other hand, in the homogeneous ordered collective phase of the Vicsek model exhibits giant fluctuations with exponent $\alpha \sim 0.8$ for both two dimensions (2D) and three dimensions (3D). Data taken from [1]

This can be understood by imagining the wine bottle potential. In Fig. 2.6, the state of the whole system is represented by a red ball. With the parameter values η and ρ_0 inside the homogeneous ordered phase in the phase diagram in Fig. 2.3, the state of the system at the beginning of numerical simulations can be regarded as an unstable state with rotational symmetry (Fig. 2.6a). As time goes by, the particles get aligned and global order eventually develops, which corresponds to the red ball falling down the slope toward the valley of the wine bottle potential. When the global order has completely developed and the system reaches a steady state, it corresponds to the red ball settled at the bottom of the potential. After that, when the ball is subjected to even infinitesimal perturbations, the ball can freely move around along the potential minimum at the bottom of the potential without any restoring force. Such motion of the ball corresponds to the NG mode. If we interpret this in the context of collective motion, the orientation of the global order corresponds to the azimuthal position of the ball from the center of the potential. Hence, just like the position of the ball fluctuates, the orientation of the collective phase does fluctuate.

Slow and long-wavelength fluctuations of orientation arising from the NG mode then couple with the density field and increase the density fluctuations. Fluctuations of the orientation field, or the coarse-grained velocity field v, in the ordered phase exist, which correspond to splay instability often discussed in liquid crystals. If such fluctuations, $\text{div} v < 0$ or $\text{div} v > 0$, are present at a certain time, in the next time step the coarse-grained density field ρ either increase or decrease in the comoving frame of the flock respectively, as shown in Fig. 2.7. Mathematically, it can be understood from the continuity equation of the density field which comes from the number conservation of the particles in the system,

$$\frac{\partial \rho}{\partial t} + \nabla \cdot (v \rho) = 0. \tag{2.5}$$

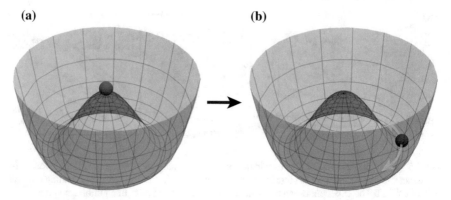

Fig. 2.6 Spontaneous breaking of rotational symmetry and the associated Nambu-Goldstone mode. Red ball represents the state of the whole system. **a** Before the symmetry breaking. The system has rotational symmetry and there is no preferred direction. **b** The symmetry broken state. Infinitesimal perturbations can drive the red ball to move around in the valley of the wine bottle potential without any restoring force. This is the Nambu-Goldstone mode

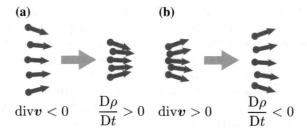

Fig. 2.7 Schematic figures of the crucial coupling between orientation and density in the collective motion. Splay instability in the orientation field $\mathrm{div}\boldsymbol{v} < 0$ or $\mathrm{div}\boldsymbol{v} > 0$ leads to the increase $\frac{\mathrm{D}\rho}{\mathrm{D}t} > 0$ or the decrease $\frac{\mathrm{D}\rho}{\mathrm{D}t} < 0$ of the density field ρ in the comoving frame of the flock. Here the derivative $\frac{\mathrm{D}}{\mathrm{D}t}$ is the Langrangian derivative defined as $\frac{\mathrm{D}}{\mathrm{D}t} = \frac{\partial}{\partial t} + \boldsymbol{v} \cdot \nabla$

This can be rewritten by introducing the Lagrangian derivative $\frac{\mathrm{D}}{\mathrm{D}t} = \frac{\partial}{\partial t} + \boldsymbol{v} \cdot \nabla$ as,

$$\frac{\mathrm{D}\rho}{\mathrm{D}t} = -\rho\,\mathrm{div}\boldsymbol{v}, \tag{2.6}$$

which clearly demonstrates the coupling between the orientational fluctuations and the density fluctuations.

In summary, the orientation field of the particles in the ordered phase with broken rotational symmetry can fluctuate due to the absence of restoring forces (the NG mode), and then these orientational fluctuations increase the density fluctuations. As a result, we observe GNF in the homogeneous ordered phase of the Vicsek model.

Hence it is very important to recognize that the GNF in the Vicsek model need to be discussed in the homogeneous ordered phase *without any clusters*, because the existence of clusters trivially increase the density fluctuations which are not the manifestation of the NG mode in the symmetry broken state.

We note that the NG mode causes algebraic correlations of the orientation field and the density field of the Vicsek model, which mean the ordered state is scale-free and thus has no clusters with any characteristic length scale. These algebraic correlations are closely related to GNF and the value of the exponent α. These correlations and their relations were calculated analytically in the hydrodynamic theories, but numerically they were quite difficult to obtain due to the limited system size.

Superdiffusion

Each particle in the ordered phase exhibits anomalous diffusion in the transverse direction of the global order. Of course, it is natural that the particles exhibit superdiffusive behavior due to their motility, but the point here is that, even if we subtract the mean velocity of the flock, each particle still shows superdiffusive behavior in the direction transverse to the mean velocity of the flock.

To evaluate the behavior of each particle, let us consider mean square displacement (MSD) in the transverse direction,

$$\Delta r_\perp^2 := \left\langle [r_\perp(t) - r_\perp(0)]^2 \right\rangle_{\text{ensemble}}, \tag{2.7}$$

where $r_\perp(t)$ denotes transverse components of the position of each particle at time t. This MSD $\Delta r_\perp^2$ scales in the long-time limit as $\Delta r_\perp^2 \sim t^\nu$ with the exponent $\nu = 4/3$ in the ordered phase of the 2D Vicsek model [1]. This exponent is also derived from the hydrodynamic theory that we will describe in the next section [4, 5, 7, 10]. Note that $\nu = 1$ in normal diffusion processes like Brownian motion.

Numerically, it is quite difficult to extract this exponent by directly calculating $r_\perp(t)$. Although we need to track particles for a very long time to see asymptotic behavior, temporal fluctuations of the direction of the global order complicate this procedure. Instead of such direct calculation, the doubling time τ_2 of the separation distance between a pair of particles in the transverse direction $\delta_\perp$ was calculated (Fig. 2.8). We can extract ν from the relation,

$$\tau_2 \sim \delta_\perp^{2/\nu}. \tag{2.8}$$

We note that this superdiffusive behavior is quite difficult to detect even in numerical simulations with periodic boundary conditions in which we can track particles in principle as long as we want. Therefore, detecting this superdiffusion in experiments would be a very hard task.

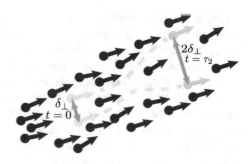

Fig. 2.8 Schematic figure on how to detect superdiffusive behavior numerically. We can estimate the exponent of superdiffusion in the transverse direction by calculating the duration τ_2 required to double the separation distance $\delta_\perp$ of a pair of particles in the ordered phase

2.3   Toner-Tu Theory

2.3.1   Idea and Definition

Although the Vicsek model demonstrates fascinating phenomena, it is a numerical model and hard to obtain any concrete analytical predictions. Therefore, John Toner and Yuhai Tu devised a hydrodynamic theory of polar flocks as seen in the Vicsek model in their pioneering papers [4–6], which is now referred to as 'the Toner-Tu theory'.

What we are interested in is the macroscopic behavior of flocks, not microscopic variables such as all the positions and the velocities of the particles. The macroscopic behavior of the system is captured by looking at the slow and long-wavelength variations of physical quantities. Such quantities, called 'hydrodynamic variables', in the Vicsek model are the coarse-grained velocity field $\boldsymbol{v}(\boldsymbol{r}, t)$ and the coarse-grained density field $\rho(\boldsymbol{r}, t)$, where $\boldsymbol{r}$ is the spatial coordinates and t is the time.

Just by writing down the terms allowed by the symmetry of the Vicsek model up to their lowest-order derivatives, Toner and Tu constructed phenomenological hydrodynamic equations for polar flocks in d-dimensional space, which is similar to the Navier-Stokes equations, as,

$$\partial_t \boldsymbol{v} + \lambda_1 (\boldsymbol{v} \cdot \boldsymbol{\nabla}) \boldsymbol{v} + \lambda_2 (\boldsymbol{\nabla} \cdot \boldsymbol{v}) \boldsymbol{v} + \lambda_3 \boldsymbol{\nabla}(|\boldsymbol{v}|^2)$$
$$= \alpha \boldsymbol{v} - \beta |\boldsymbol{v}|^2 \boldsymbol{v} - \boldsymbol{\nabla} P_1 - \boldsymbol{v}(\boldsymbol{v} \cdot \boldsymbol{\nabla} P_2) + D_B \boldsymbol{\nabla}(\boldsymbol{\nabla} \cdot \boldsymbol{v}) + D_T \nabla^2 \boldsymbol{v} + D_2 (\boldsymbol{v} \cdot \boldsymbol{\nabla})^2 \boldsymbol{v} + \boldsymbol{f}, \tag{2.9}$$

$$\partial_t \rho + \boldsymbol{\nabla} \cdot (\boldsymbol{v}\rho) = 0, \tag{2.10}$$

$$P_1 = \sum_{n=1}^{\infty} \sigma_n(|\boldsymbol{v}|)(\rho - \rho_0)^n, \tag{2.11}$$

$$P_2 = \sum_{n=1}^{\infty} \mu_n(|\boldsymbol{v}|)(\rho - \rho_0)^n, \tag{2.12}$$

where λ_1, λ_2, λ_3, α, β, P_1, and P_2 are, in general, functions of the local density ρ and the magnitude of the local velocity $|\boldsymbol{v}|$, and Eqs. (2.11) and (2.12) are corresponding Taylor expansions. P_1 is the isotropic pressure term and P_2 is the anisotropic pressure term. β, D_B, D_T, D_2 are all positive. The D_B, D_T, D_2 terms represent diffusion of fluctuations of the velocity field via interactions among particles. The terms $\alpha\boldsymbol{v} - \beta|\boldsymbol{v}|^2\boldsymbol{v}$ represent a Ginzburg-Landau type potential $-\frac{1}{2}\alpha|\boldsymbol{v}|^2 + \frac{1}{4}\beta|\boldsymbol{v}|^4$ that leads to spontaneous symmetry breaking depending on the sign of α. Negative and positive values of α correspond to the disordered phase and the ordered phase, and hence we can describe the order-disorder transition with these terms. The ordered phase has non-zero self-propulsion speed $v_0 = \sqrt{\frac{\alpha}{\beta}}$. The $\boldsymbol{f}$ term is a white Gaussian noise with delta correlations,

$$\langle f_i(\boldsymbol{r}, t) f_j(\boldsymbol{r}', t)\rangle = \Delta\delta_{ij}\delta^d(\boldsymbol{r} - \boldsymbol{r}')\delta(t - t'), \qquad (2.13)$$

where Δ is a constant, and i and j denote Cartesian components.

The λ terms in the left-hand side of Eq. (2.9) correspond to the advection terms, and are allowed due to the absence of Galilean invariance. In usual hydrodynamic equations like the Navier-Stokes equations, Galilean invariance of the system requires the total momentum of the system to be conserved. In accordance with this requirement, such λ terms are not allowed and the values of λ_1, λ_2, and λ_3 have to be $\lambda_1 = 1$ and $\lambda_2 = \lambda_3 = 0$. However, in systems such as the Vicsek model, bird flocks, fish schools, and mammalian herds we consider here, there is no Galilean invariance and the total momentum of the system is *not* conserved because there exist single absolute frames of reference in which the surrounding environment such as air, water, the ground is at rest and we just look at the dynamics of each particle separated from the surroundings. In a sense, we assume implicitly the existence of s ome sort of invisible 'momentum sinks' and neglect their dynamics. For example, dynamics of water is often neglected when looking at schools of fish or bacterial suspensions, and that of air is also neglected when looking at flocks of birds. We usually do not consider momentum of the earth or the substrate when we think about mammalian herds and bacterial swarms on agar plates, and here the earth and the agar really work as momentum sinks. Such active matter systems without momentum conservation are called 'dry' active matter systems, and the systems in which we explicitly consider surrounding fluids and the total momentum is conserved are called 'wet' active matter systems such as bulk bacterial suspensions.

The anisotropic pressure term P_2 in Eq. (2.9) is allowed due to nonequilibrium nature of polar flocks and such terms cannot appear in equilibrium systems due to Pascal's law. When the Toner-Tu equations were proposed in 1995 [4, 5, 7], this term had been overlooked. In 2012, Toner incorporated this term and reanalyzed the equations. However, it turned out that this term complicates their dynamical renormalization group analysis on their equations because the additional nonlinear terms emerging from this anisotropic pressure term are relevant under a dynamical renormalization group.

Although some of the results obtained in their previous works [4, 5, 7, 10] were invalidated by the new analysis with the P_2 term [6], some other important results are still undoubtedly valid. Furthermore, there is no other analytical theory for polar flocks. So it is still worthwhile to review their results and compare experimental results with these Toner-Tu results in order to gain insight on collective motion.

Finally, we remark that the value of coefficients in such hydrodynamic equations cannot be determined solely by symmetry arguments. These hydrodynamic equations can be derived by applying the Boltzmann equation approach to microscopic Vicsek-style models, and then we can obtain the representations of coefficients in hydrodynamic equations in terms of microscopic variables [11–13]. However, the Boltzmann approach relies on some assumptions such as molecular chaos and dilute limit to ensure binary collisions, so such representations may not be quantitatively precise.

2.3.2 Predictions from Toner-Tu Theory

True long-range order

The most important and rigorous result from the Toner-Tu theory is that the polar flocks considered there have true long-range order (true LRO). Because the Toner-Tu theory is based on just symmetry arguments on the Vicsek model, its results should also be applied to microscopic simulations on the Vicsek model. The result on LRO is unchanged after the reanalysis that includes the P_2 term.

The presence of true LRO in the Vicsek model is a nonequilibrium and nonlinear effect. The Vicsek model exhibits true long-range order even in $d = 2$, which is impossible in equilibrium systems. In equilibrium, 2D systems with short-ranged interactions such as the classical XY model and nematic liquid crystals cannot break continuous symmetry, which is stated in the Mermin-Wagner theorem [14]. The long-wavelength fluctuations originating from the NG mode cannot be suppressed and they cannot achieve true long-range order, resulting in quasi-long-range order with many topological defects.[5] Therefore, what we can observe in such 2D equilibrium systems with rotational symmetry is not phase transitions associated with broken symmetry but a topological phase transition called 'Kosterlitz-Thouless transiton' [15].

Intuitive explanation for true long-range order

The existence of true LRO in the Vicsek model can be understood by thinking about spreading of information on orientations. Intuitively, neighboring particles of each particle in the Vicsek model are continuously changing due to orientational fluctuations. Therefore, particles in the Vicsek model can interact more particles than

---

[5]True and quasi-long-range order can be distinguished by correlation functions. If the correlation remains positive at the large system size limit, the system has true long-range order. On the other hand, the correlation decays algebraically in the case of quasi-long-range order.

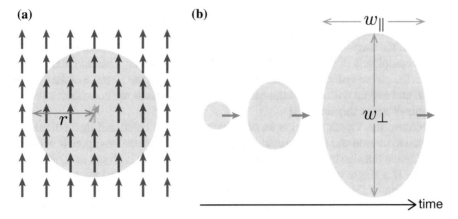

Fig. 2.9 Schematic figures on how the information of orientational fluctuations spreads inside **a** the XY model on a lattice and **b** the polar flocks of the Vicsek model. Pink regions represent the areas in which the orientational information has spread out. In the XY model, information spreads isotropically. On the other hand, in the off-lattice Vicsek model, information spreads more rapidly in the transverse direction than in the longitudinal direction due to the advection of particles

fixed particles, and the interaction radius can be effectively much larger than that of the equilibrium models such as the classical XY model in which spins are fixed on lattices. Here we follow the argument introduced by John Toner and published in Francesco Ginelli's lecture note [9].

To look at the difference between the equilibrium XY model and the Vicsek model, let us first examine the XY model on a d-dimensional lattice. Suppose all the spins are directing in exactly the same direction except for one single spin. Our questions are the following: What happens if this spin deviates from the direction of the other spin by an angle $\delta\theta_0$ as shown in Fig. 2.9a? How does the information on this fluctuation, or error, spread among the lattice? On this lattice, such a deviation propagates just by diffusion. Hence after a certain period of time τ, this error diffuses over a distance $r \sim \sqrt{\tau}$, or a volume $V \sim r^d \sim \tau^{d/2}$. If we assume that the total error inside this volume is conserved, the error per spin $\delta\theta$ scales as $\delta\theta \sim \delta\theta_0/V \sim \delta\theta_0\tau^{-d/2}$. So far we have calculated information spreading of a single error, but in reality, there exist multiple errors whose numbers should be proportional to the duration and the volume of interest. Hence in the volume V we have $n_e \sim \tau V$ errors. Because errors can be both positive and negative and they are subject to the central limit theorem, the total amplitude of the noise Ω can be estimated as $\Omega \sim \sqrt{n_e} \sim \sqrt{\tau V}$. Therefore, the total error amplitude per spin $\Delta\theta$ can be estimated as,

$$\Delta\theta \sim \frac{\Omega}{V} \sim \sqrt{\frac{\tau}{V}} \sim r^{1-d/2} \rightarrow \begin{cases} 0 & (d > 2) \\ \ln L \rightarrow \infty & (d = 2) \\ \infty & (d < 2) \end{cases} \quad (2.14)$$

This means that in $d > 2$ such orientational errors are damped algebraically and the global order is resistant to such fluctuations, resulting in true long-range order. In $d < 2$, the fluctuations grow algebraically in space and the global order present at first is completely destroyed, resulting in no long-range order. The marginal case in $d = 2$, the global order is destroyed extremely slowly with the increase of the system size L and end up with quasi-long-range order. These results are consistent with the Mermin-Wagner theorem [14].

Because the Vicsek model is an off-lattice model, orientational fluctuations are coupled to motion and the information on orientational fluctuations spreads not only by diffusion but also by particle motion, or advection, which is absent in equilibrium models. If a single particle in the Vicsek model has orientational fluctuations $\Delta\theta$, the position of this particle deviates from the global mean direction of motion by $\delta x_\perp \sim v_0\tau \sin\Delta\theta \sim \tau\Delta\theta$ in the transversal direction and $\delta x_\parallel \sim v_0\tau(1 - \cos\Delta\theta) \sim \tau\Delta\theta^2$ in the longitudinal direction. Due to the anisotropy of polar flocks in the Vicsek model, it is natural to decompose the volume V in which the information on error has spread out as $V \sim w_\perp{}^{d-1}w_\parallel$ as depicted in Fig. 2.9b. From the same argument as in the equilibrium XY model, $\Delta\theta$ is given by,

$$\Delta\theta \sim \sqrt{\frac{\tau}{V}} = \frac{\tau^{1/2}}{\sqrt{w_\perp{}^{d-1}w_\parallel}} \sim \tau^\gamma, \tag{2.15}$$

where we defined an exponent γ. The transversal and the longitudinal length scales, $w_\perp$ and $w_\parallel$, over which the information on error has reached are estimated in terms of advection and diffusion,

$$w_\perp \sim \delta x_\perp + D_\perp \tau^{1/2} \sim \tau\Delta\theta + D_\perp\tau^{1/2} \sim \tau^{\gamma_\perp}, \tag{2.16}$$

$$w_\parallel \sim \delta x_\parallel + D_\parallel \tau^{1/2} \sim \tau\Delta\theta^2 + D_\parallel\tau^{1/2} \sim \tau^{\gamma_\parallel}, \tag{2.17}$$

where $D_\parallel$ and $D_\perp$ are diffusion constants, and we defined the exponents γ, $\gamma_\perp$, and $\gamma_\parallel$ so that $\gamma_\perp = \max(1 + \gamma, 1/2)$ and $\gamma_\parallel = \max(1 + 2\gamma, 1/2)$. By inserting Eqs. (2.16) and (2.17) into Eq. (2.15), we obtain,

$$\gamma = \frac{1}{2} - \frac{d}{4}, \quad \gamma_\perp = \gamma_\parallel = \frac{1}{2} \quad \text{for} \quad d \geq 4, \tag{2.18}$$

$$\gamma = \frac{3 - 2d}{2(d+1)}, \quad \gamma_\perp = \frac{5}{2(d+1)}, \quad \gamma_\parallel = \frac{1}{2} \quad \text{for} \quad \frac{7}{3} \leq d < 4, \tag{2.19}$$

$$\gamma = \frac{1-d}{d+3}, \quad \gamma_\perp = \frac{4}{d+3}, \quad \gamma_\parallel = \frac{5-d}{d+3} \quad \text{for} \quad d < \frac{7}{3}, \tag{2.20}$$

all of which give negative γ for any $d > 1$. This means that orientational fluctuations are suppressed and the Vicsek model can possess true long-range order at any dimension $d > 1$. Although, of course, the arguments presented here are not precise compared with dynamical renormalization group analysis done by Toner and Tu [4, 5, 7], this captures intuitive understanding on how and why true long-range order is possible in the Vicsek model. The coupling between positions and orientational fluctuations leads to faster information spreading of orientational fluctuations via advection.

Correlation functions

Toner and Tu defined correlation functions on the velocity field and the density field in the ordered phase. Their asymptotic behaviors were calculated by dynamical renormalization group analysis [4, 5]. These results were later invalidated by the P_2 term [6].

Definitions of the correlation functions are the followings.[6]

$$C(\boldsymbol{R}) := \langle \boldsymbol{v}(\boldsymbol{r} + \boldsymbol{R}, t) \cdot \boldsymbol{v}(\boldsymbol{r}, t) \rangle, \tag{2.21}$$

$$C_C(\boldsymbol{R}) := \langle \boldsymbol{v}(\boldsymbol{r} + \boldsymbol{R}, t) \cdot \boldsymbol{v}(\boldsymbol{r}, t) \rangle - |\langle \boldsymbol{v} \rangle|^2 = \langle \boldsymbol{v}_\perp(\boldsymbol{r} + \boldsymbol{R}, t) \cdot \boldsymbol{v}_\perp(\boldsymbol{r}, t) \rangle, \tag{2.22}$$

$$C_{ij}(\boldsymbol{q}, \omega) := \langle v_i^\perp(\boldsymbol{q}, \omega) v_j^\perp(-\boldsymbol{q}, \omega) \rangle, \tag{2.23}$$

$$C_{ij}(\boldsymbol{q}) := \langle v_i^\perp(\boldsymbol{q}, t) v_j^\perp(-\boldsymbol{q}, t) \rangle, \tag{2.24}$$

$$C_\rho(\boldsymbol{q}) := \langle |\rho(\boldsymbol{q}, t)|^2 \rangle = \langle \rho(\boldsymbol{q}, t) \rho(-\boldsymbol{q}, t) \rangle, \tag{2.25}$$

where average $\langle \ \rangle$ is taken over space, time or ensembles, $\boldsymbol{v}_\perp$ is a velocity component perpendicular to the direction of the global order, $v_i^\perp$ is the i-direction component of $\boldsymbol{v}_\perp$. $\boldsymbol{q}$ is a wave vector in the Fourier space. We note that the equal-time density correlation function in the Fourier space $C_\rho(\boldsymbol{q})$ defined in Eq. (2.25) is also called the structure factor $S(\boldsymbol{q})$.

The exponents of the asymptotic behaviors of correlation functions are defined as follows:

$$C_C(\boldsymbol{R}) \propto R_\perp^{2\chi}, \tag{2.26}$$

$$C_C(\boldsymbol{R}) \propto R_\parallel^{2\chi/\zeta}, \tag{2.27}$$

where the subscript $\perp$ and $\parallel$ denote transverse and longitudinal components with respect to the direction of the global order respectively. The exponent ζ is a measure of anisotropy of the ordered phase.

Their dynamical renormalization group analysis gives asymptotic behaviors of other correlation functions in terms of the exponents defined above,

---

[6]In their original papers [4, 5, 7], there are 'fluctuations of notation', which we have corrected in this thesis.

$$C_\rho(\boldsymbol{q}) = S(\boldsymbol{q}) = \langle|\rho(\boldsymbol{q},t)|^2\rangle \sim \begin{cases} q_\perp^{1-d-\zeta-2\chi} & (q_\parallel \ll q_\perp) \\ q_\parallel^{-2}q_\perp^{3-d-\zeta-2\chi} & (q_\perp \ll q_\parallel \ll q_\perp^\zeta) \\ q_\parallel^{-3+(1-d-2\chi)/\zeta}q_\perp^2 & (q_\perp^\zeta \ll q_\parallel) \end{cases} \quad (2.28)$$

$$C_{ij}(\boldsymbol{q}) = \langle v_i^\perp(\boldsymbol{q},t)v_j^\perp(-\boldsymbol{q},t)\rangle \sim q_\perp^{1-d-\zeta-2\chi} \tag{2.29}$$

$$\Delta N \propto \langle N\rangle^\alpha = \langle N\rangle^{\frac{1}{2}+\frac{-1+d+\zeta+2\chi}{2d}} \tag{2.30}$$

Note that $C_{ij}(\boldsymbol{q})$ can be calculated from $C_C(\boldsymbol{R})$ from the following relation,

$$C_C(\boldsymbol{R}) = \langle \boldsymbol{v}_\perp(\boldsymbol{r}+\boldsymbol{R},t)\cdot\boldsymbol{v}_\perp(\boldsymbol{r},t)\rangle = \int\frac{d^d\boldsymbol{q}}{(2\pi)^d}C_{ii}(\boldsymbol{q})e^{i\boldsymbol{q}\cdot\boldsymbol{R}}. \tag{2.31}$$

Giant number fluctuations

Here we estimate the exponent of the number fluctuations α defined as,

$$\Delta N \propto \langle N\rangle^\alpha. \tag{2.32}$$

Number fluctuations are generally related to the structure factor as,

$$S(\boldsymbol{q}\to\boldsymbol{0}) = \rho_0\left[\frac{\Delta N^2}{\langle N\rangle}\right]_{N\to\infty} \sim \langle N\rangle^{2\alpha-1}. \tag{2.33}$$

From the Eq. (2.28), we can see that the divergence of $C_\rho(\boldsymbol{q}) = S(\boldsymbol{q})$ when $\boldsymbol{q}\to\boldsymbol{0}$ is dominated in the $q_\parallel \ll q_\perp$ region, so we can write,

$$S(\boldsymbol{q}\to\boldsymbol{0}) \sim \frac{1}{q^\sigma} \sim l^\sigma \sim \langle N\rangle^{\sigma/d} \quad \text{with} \quad \sigma = -(1-d-\zeta-2\chi), \tag{2.34}$$

where l is a typical length scale.

Then, by equating the exponents in Eqs. (2.33) and (2.34), we obtain

$$\alpha = \frac{1}{2}+\frac{\sigma}{2d} = \frac{1}{2}+\frac{-(1-d-\zeta-2\chi)}{2d}. \tag{2.35}$$

Hence the exponent α is larger than $1/2$, so the Toner-Tu theory also predicts giant number fluctuations in the ordered phase.

Summary on predictions for $d = 2$

Only in $d = 2$, they obtained the exact values of the exponents. This is because in $d = 2$ the term $\lambda_2(\boldsymbol{\nabla}\cdot\boldsymbol{v})\boldsymbol{v}$ in Eq. (2.9) become equivalent to the term $\lambda_1(\boldsymbol{v}\cdot\boldsymbol{\nabla})\boldsymbol{v}$ when we rewrite Eq. (2.9) by inserting $\boldsymbol{v} = v_0\hat{\boldsymbol{e}}_\parallel + \delta\boldsymbol{v}$ and neglecting irrelevant terms, where $\hat{\boldsymbol{e}}_\parallel$ is a unit vector along the global order.

The theoretical prediction for the exponents in $d = 2$ is $\chi = -\frac{1}{5}$, $\zeta = \frac{3}{5}$, and $\alpha = \frac{4}{5}$. Therefore, all the predictions for the exponents in $d = 2$ by the Toner-Tu theory are the following:

$$C_C(R) = \langle v_\perp(r + R, t) \cdot v_\perp(r, t) \rangle \propto \begin{cases} R_\perp^{2\chi} = R_\perp^{-2/5} = R_\perp^{-0.4} \\ R_\parallel^{2\chi/\zeta} = R_\parallel^{-2/3} = R_\parallel^{-0.667}, \end{cases} \quad (2.36)$$

$$C_{ij}(q) = \langle v_i^\perp(q, t) v_j^\perp(-q, t) \rangle \sim q_\perp^{1-d-\zeta-2\chi} = q_\perp^{-6/5} = q_\perp^{-1.2}, \quad (2.37)$$

$$C_\rho(q) = S(q) = \langle |\rho(q, t)|^2 \rangle \sim q_\perp^{1-d-\zeta-2\chi} = q_\perp^{-6/5} = q_\perp^{-1.2}, \quad (2.38)$$

$$\Delta N \propto \langle N \rangle^\alpha = \langle N \rangle^{\frac{1}{2} + \frac{-(1-d-\zeta-2\chi)}{2d}} = \langle N \rangle^{4/5} = \langle N \rangle^{0.8}. \quad (2.39)$$

We note that the exponent for GNF $\alpha = 0.8$ obtained here is consistent with the numerical results of the Vicsek model.

2.3.3 Remarks for Analysis with a Finite System Size

Except for the exponent α for number fluctuations, the other exponents in correlation functions are hard to calculate mainly due to the limited system sizes of simulations. In the Toner-Tu theory, we can treat infinitely large space and can exactly define the direction of the global order and the mean velocity. On the other hand, in numerical studies we have to estimate them from finite system sizes and need to subtract the mean velocity from the velocity field for obtaining fluctuations of the velocity field $v_\perp$ and calculating $C_C(R) = \langle v_\perp(r + R, t) \cdot v_\perp(r, t) \rangle$ and $C_{ij}(q) = \langle v_i^\perp(q, t) v_j^\perp(-q, t) \rangle$. Because $v_\perp$ and $v_i^\perp$ are deviations from the global mean velocity, there are constraints: $\langle v_\perp(r, t) \rangle_r = 0$ and $\langle v_i^\perp(r, t) \rangle_r = 0$. Hence correlation functions of fluctuations in the real space such as $C_C(R)$ have to become negative at a certain distance away from the origin $R = 0$. The same problem occurs in experiments because we have to estimate the direction of the global order from finite observation areas.

2.3.4 Toner-Tu-Ramaswamy Phase

In the above subsections, we have examined the properties of the ordered state in the Toner-Tu theory. This broken symmetry state has true long-range order even in $d = 2$ and exhibits algebraic correlations associated with the Nambu-Goldstone mode both in the velocity field and in the density field. Because the correlations are algebraic and scale-free, there are no clusters with any characteristic length scale and this phase is spatially homogeneous. However, both the density field and the velocity field is highly fluctuating, which results in the giant number fluctuations. All of these properties are arising from nonlinear nonequilibrium nature of the system and the Nambu-Goldstone mode in this broken symmetry state.

This state is now often referred to as 'the Toner-Tu phase'. However, GNF were originally predicted in the hydrodynamic theory for collective motion of apolar particles with nematic interactions (active nematics) by Sriram Ramaswamy et al. [8], so we call this phase 'the Toner-Tu-Ramaswamy phase (the TTR phase)'.

2.4 Vicsek-Style Models with Different Symmetry

2.4.1 Definitions on Active Nematics and Self-propelled Rods

So far we have reviewed the numerical results and the hydrodynamic theories on the original Vicsek model, which deals with collective motion of polar particles with ferromagnetic interactions. As we have already mentioned in the beginning of this chapter, we can also consider Vicsek-style models with other symmetries on motility and interactions. Here, 'Vicsek-style' means that (i) particles are point-like and (ii) their interactions are local, short-ranged, and calculated by averaging the directions of neighbors within the interaction radius. By considering these models, we can gain a clearer view both on how the true long-range order emerges in the Vicsek model and on the universality of these simple models.

The two Vicsek-style models we consider here are called 'active nematics' [16, 17] and 'self-propelled rods' [18], both of which have nematic interactions but different motilities. Dynamics of each particle in these models is given as follows.

active nematics

$$\theta_j^{t+1} = \frac{1}{2} \arg \sum_{k \sim j} e^{2i\theta_k^t} + \eta_j^t, \tag{2.40}$$

$$r_j^{t+1} = r_j^t \pm v_0 e_{\theta_j^{t+1}}, \tag{2.41}$$

where the sign $\pm$ in Eq. (2.41) is chosen at $1/2$ probabilities at every time step.

Note that θ_k^t inside the exponential function in Eq. (2.40) is multiplied by 2, which represents nematic interactions.

self-propelled rods

$$\theta_j^{t+1} = \arg \sum_{k \sim j} \text{sign}\left[\cos(\theta_k^t - \theta_j^t)\right] e^{i\theta_k^t} + \eta_j^t, \tag{2.42}$$

$$r_j^{t+1} = r_j^t + v_0 e_{\theta_j^{t+1}}. \tag{2.43}$$

In self-propelled rods systems, each particle moves persistently toward its head direction without directional reversal.

These two models are pure modifications of the Vicsek model, because all of the models have in common that particles are point-like and their interactions are calculated by averaging the directions of neighbors within the interaction radius. Hence these variant models are called Vicsek-style models.

Particles in the active nematics systems have head-tail symmetry, and they are apolar. They stochastically move back and forth, and interact with each other nematically. This model corresponds to experimental systems of collection of symmetric rods vibrated on a plate [19]. Particles in self-propelled rods imitate some kinds of bacteria swarming on agar plates as depicted in Fig. 2.10a [20]. They align in either parallel or antiparallel configurations.

The interaction terms in Eqs. (2.40) and (2.42) are unchanged under the transformations $\theta_k^t \rightarrow \theta_k^t + \pi$ or $\theta_j^t \rightarrow \theta_j^t + \pi$. These π-symmetric interactions represent nematic interactions.

(a)   **(b)**

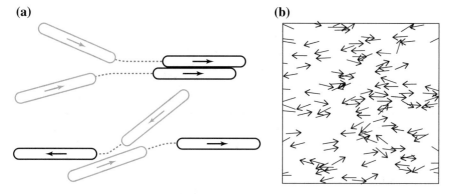

Fig. 2.10 **a** Nematic interactions between self-propelled rods. They can align both in parallel and antiparallel configurations. **b** The ordered phase of self-propelled rods model. It exhibits nematic order

2.4.2 Properties of the Models

Let us look at the properties of active nematics and self-propelled rods in comparison with the original Vicsek model. All the properties are summarized in Fig. 2.11.

First-order phase transition

Both of the two models exhibit discontinuous first-order phase transitions. There exist coexistence phases, and ordered regions emerge as band structures inside the disordered regions. Although in the Vicsek model the band structures, or the Vicsek wave, stably propagate in the system, band structures in active nematics and self-propelled rods exhibit spatiotemporally chaotic behavior by deforming, splitting, and merging irregularly.

Giant number fluctuations

GNF are obtained in the ordered phases both in active nematics [17] and self-propelled rods [18]. Numerical studies have shown that the GNF exponent α is around 0.8 in both models, and α is slightly smaller than 0.8 in self-propelled rods [18]. Although corresponding hydrodynamic descriptions for $d = 2$ are devised/derived for both active nematics [8, 12] and self-propelled rods [11] as the Toner-Tu equations for the Vicsek model, *no exact results for the exponents have been obtained.* This is because these systems have another hydrodynamic variable other than the velocity field v and the density field ρ, which is the tensor order parameter Q used for nematic liquid crystals. Q is defined by using unit orientation vectors a of particles as $Q_{ij} = \langle a_i a_j \rangle_l - \frac{1}{d}\delta_{ij}$, where $\langle\ \rangle_l$ is a local average. This additional variable Q complicates the dynamics and makes it difficult to obtain analytical results.

	Vicsek model	active nematics	self-propelled rods
schematics			
motility	polar	apolar	polar
interaction	ferromagnetic	nematic	nematic
$\Delta N \propto \langle N \rangle^{\alpha}$ (numerics)	$\alpha \sim 0.8$	$\alpha \sim 0.8$	$\alpha \sim 0.75$
$\Delta N \propto \langle N \rangle^{\alpha}$ (continuum theory)	$\alpha = 0.8$ (d=2, Toner-Tu)	$\alpha = 1$ (linear theory)	no calculation
order	true long-range	quasi-long-range	true long-range

Fig. 2.11 Classification of three basic Vicsek-style models: the original Vicsek model, active nematics, and self-propelled rods. These models can be classified according to polarities of motility and interactions of particles. All the models exhibit giant number fluctuations with the exponent α around 0.8 in their homogeneous ordered states, or the Toner-Tu-Ramaswamy phases

The hydrodynamic description for active nematics is much simpler than that for self-propelled rods and some analytical calculations have been performed, because the velocity field v relaxes quite rapidly to 0 due to apolar motility of each particle and v does not play any role at the macroscopic hydrodynamic level. So the hydrodynamic equations for active nematics are written by two variables ρ and $\boldsymbol{w} = \rho \boldsymbol{Q}$ as [12, 21],

$$\frac{\partial \rho}{\partial t} = \frac{1}{2}\nabla^2 \rho + \frac{1}{2}\boldsymbol{\Gamma} : \boldsymbol{w}, \tag{2.44}$$

$$\frac{\partial \boldsymbol{w}}{\partial t} = \mu \boldsymbol{w} - 2\xi \boldsymbol{w}(\boldsymbol{w} : \boldsymbol{w}) + \frac{1}{2}\nabla^2 \boldsymbol{w} + \frac{1}{8}\boldsymbol{\Gamma}\rho, \tag{2.45}$$

where μ and ξ are coefficients, and $\boldsymbol{\Gamma}$ is a tensor differential operator with components $\Gamma_{11} = -\Gamma_{22} = \partial_1 \partial_1 - \partial_2 \partial_2$ and $\Gamma_{12} = \Gamma_{21} = 2\partial_1 \partial_2$.

Ramaswamy et al. calculated correlation functions using *linearized* equations of Eqs. (2.44) and (2.45), and accordingly discovered the existence of giant number fluctuations in the ordered phase for the first time. This linear theory predicts the exponent $\alpha = 1$, or $\Delta N \propto \langle N \rangle^1$ [8]. However, $\alpha \leq 1$ by definition and therefore this estimate by the linear theory corresponds to the upper bound.

At the nonlinear level, a perturbative renormalization group treatment has been performed for active nematics *without the density field*, and concluded that the linear predictions should hold [22]. However, as already pointed out by themselves in [7, 22], nonlinear effects, especially some involving the density field, could change all of the above renormalization group arguments. Therefore, no exact values for the exponents of correlation functions and GNF in the case of active nematics as well as self-propelled rods are known.

True versus Quasi long-range order

The crucial difference in these models is whether they have true long-range order or not. Active nematics do not have true long-range order, which has been confirmed by numerical simulations [16, 17, 21] and both the linear and the nonlinear hydrodynamic theories [22]. On the other hand, self-propelled rods exhibit true long-range order at least numerically. However, due to the complexity of the hydrodynamic equations for self-propelled rods, there is still no analytical calculation on whether self-propelled rods can really show true long-range order or not.

In the cases of active nematics and self-propelled rods, whether their nematic phases are true long-ranged or not can be evaluated by calculating a scalar nematic order parameter $S = \frac{1}{N}\left|\sum_{j=1}^{N} e^{2i\theta_j^t}\right|$, or its time-averaged value $\langle S \rangle_t$, as a function of the system size. In the case of active nematics, $\langle S \rangle_t$ decays algebraically with a very small exponent toward 0 in the large system size limit, which means quasi-long-range order. On the other hand, in self-propelled rods, $\langle S \rangle_t$ algebraically converges to a positive finite value in the large system size limit, which means true long-range order.

As we have seen above, the symmetry of the particles' motility changes the nature of the ordered phase: true or quasi-long-range order. However, in fact, there is a legitimate question raised in past works [23], which is whether or not self-propelled 'rods' are an entirely different class from polar flocks and active nematics. The globally nematic phase in self-propelled rods might be seen as the superposition of two polar systems exchanging particles at some rate. As remarked in [18], this rate is very low and it defines a finite but large time/length, over which particles go in one of the two main directions defining the global nematic order. This length scale requires enormously large system size simulations which are almost inaccessible. Hence analytical calculation on the hydrodynamic equations for self-propelled rods is needed as future work.

However, there is still a reasonable reason to believe the numerical result that the self-propelled rods have true long-range order. In the original Vicsek model, as we have seen before, true long-range order is a consequence of advection of orientational fluctuations arising from particles' persistent motion. In active nematics, each particle does not move persistently, so the information of orientational fluctuations is not advected by particles' motion transverse to the global order. Hence, active nematics cannot attain true long-range order, resulting in quasi-long-range order. On the other hand, in self-propelled rods systems, each particle moves persistently without directional reversals, so the information of orientational fluctuations can be advected by transversal velocity fluctuations. Therefore, we can still believe that the same mechanism for the emergence of true long-range order in the Vicsek model works for self-propelled rods.

2.5 Vicsek Universality Class

As we have seen so far, all the Vicsek-style models have many common properties: first-order phase transition, giant number fluctuations in the ordered phase, etc. Corresponding hydrodynamic descriptions, which are derived just by symmetry arguments, also reproduce such properties. The phase transitions in these models and field equations are now understood by a phase-separation scenario between a disordered 'gas' state and an orientationally-ordered 'liquid' state with a coexistence phase [24, 25]. However, unlike equilibrium liquid-gas phase transitions, there is no supercritical region in this far-from-equilibrium liquid-gas phase transition, in which liquid states and gas states cannot be distinguished.

As such, these models and field equations are considered to constitute a sort of nonequilibrium universality class, 'the Vicsek universality class', although there is still no concrete definition on this because of controversy on whether the Vicsek model, active nematics, and self-propelled rods are indeed in the same class or not. From the discussion above, sometimes we divide them into two classes according to their nature of long-range order: the Vicsek class and the active nematics class. In this case, self-propelled rods are classified into the Vicsek class in this 'narrow' definition. However, there is a consensus that the ordered phases of these models and field equations, or the Toner-Tu-Ramaswamy phases, undoubtedly have universal properties in common: giant number fluctuations in the homogeneous long-range ordered phase. Whether or not such universality can really be observed experimentally remained elusive, for which we have given the first example that will be detailed in Chap. 3 [26].

2.6 Experimental Efforts to Find the TTR Phases

2.6.1 Experimental Difficulties

Motivated by biological collective motion such as flocks of birds, schools of fish, swarms of bacteria, and herds of mammals, numerical and theoretical studies on collective motion have revealed its fundamental properties from the viewpoint of statistical physics. However, these results are restricted to 'imaginary flocks' in the models or the field equations, so as a discipline of natural science it is important to verify whether these properties are really present in real flocks.

Although the studies on collective motion were motivated by biological flocks, those flocks cannot belong to the Vicsek universality class. Biological flocks such as flocks of birds and schools of fish are spatially localized and do not have long-range order. Past theoretical works deal with spatially-homogeneous ordered collective phases that are phenomenologically different from spatially-localized flocks of birds, etc. Furthermore, it is almost impossible to conduct controlled experiments with birds or fish because we cannot demand them to interact differently or to form denser/sparser flocks. On the other hand, we need to pursue 'homogeneous long-range ordered phases' as seen in the Vicsek-style models and corresponding hydrodynamic equations.

In need of controllable systems simpler than bird flocks and fish schools, various experimental systems are devised and utilized in order to search for GNF in the TTR phases. Examples of such experiments include biofilaments driven by molecular motors [27], colloids consuming electric energy [28], shaken granular materials [19, 29, 30], monolayers of fibroblast cells [31], and common bacteria [20, 32]. However, none of these experiments has been fully convincing in demonstrating the presence of GNF *in the true sense of the Vicsek universality class* as predicted from the works of Toner et al. [4–8, 23], and observed in Vicsek-style models [3, 17, 18, 33]: In some cases, only normal number fluctuations were found [27, 28]. In others, GNF were reported for systems *not* in the fully ordered phase [19, 20, 27, 29, 31]. Finally, Refs. [30, 32] show some evidence of GNF only in numerical models of the experiments described.

Difficulties and pitfalls indeed abound to observe unambiguous Toner-Tu-Ramaswamy phenomena: Very large systems are typically needed; external boundaries thus prevent their observation; it is often difficult to distinguish the coexistence phase from the orientationally-ordered liquid phase, leading one to confuse the non-asymptotic fluctuations due to clustering with the GNF of the orientational liquid; strong steric interactions in dense systems may overcome alignment effects;

additional long-range interactions may tame density fluctuations. Due to such experimental difficulties and complicated theoretical background, there has been a misconception even among the active matter community that 'GNF' have been detected experimentally and are not such surprising/important property.

Before moving on to the discussion on each existing experiment, let us summarize and state clearly again here what we mean by the 'Toner-Tu-Ramaswamy phase': a fluctuating, orientationally-ordered phase with the sort of long-range correlations and anomalous fluctuations without phase-separation into dense clusters sitting in a disordered sparse gas. With this definition in mind, we will detail above experiments one by one in the following subsection.

2.6.2 Experimental Systems

Here we detail existing experimental studies reporting 'giant number fluctuations', which is *not* actually in the sense of Toner-Tu-Ramaswamy phenomena. In most cases, observation of 'GNF' was not the main claims of those paper, but such reports have made many people confused about the original meaning of 'the genuine GNF' predicted by Toner, Tu, Ramaswamy et al.

There are, of course, many other studies on collective motion of active matter systems, but here we introduce studies that report 'GNF'.

Shaken granular rods

Narayan, Ramaswamy et al. conducted experiments using granular rods vibrated on a plate [19] as shown in Fig. 2.12. The symmetry of the system corresponds to active nematics, and they indeed observed quasi-long-range order in correlation functions. In this quasi-long-range ordered state, they report 'GNF'. However, as is often the case for granular experiments, this experiment suffered from boundary effects and small system size. The lengths of the rods were ~4.6 mm and the diameter of the plate was 13 cm, and the number of rods was 2820 at the largest. Therefore, the rods were strongly affected by the boundary and aligned along the boundary. As a result, there exist many topological defects in the director field of the rods. The 'GNF' here were reported in such phases, so we cannot exclude the possibility that such 'GNF' are a consequence of the existence of topological defects (Fig. 2.12). We cannot conclude that the 'GNF' obtained here are originating from the Nambu-Goldstone mode of the symmetry broken TTR phase. The GNF in the sense of the Vicsek universality class should be discussed in a certainly long-range ordered phase without apparent defects, even in quasi-long-range ordered phases.

Furthermore, due to the small experimental system size, they observed the 'GNF' only over less than a decade. Extracting the exponents of algebraic behavior from such a narrow range is not a good idea.

(a) **(b)**

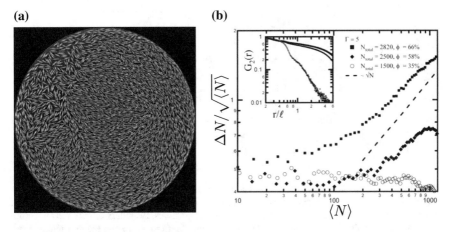

Fig. 2.12 **a** Snapshot of experiments on shaken granular rods with the number of rods $N_{\text{total}} = 2820$. The upper-left region is sparse, which leads to 'GNF'. Many topological defects exist and no real global order emerges. **b** The magnitude of normalized number fluctuations $\Delta N/\sqrt{\langle N \rangle}$ vs $\langle N \rangle$. At high density with $N_{\text{total}} \geq 2500$, they observe 'GNF'. However, it is limited only over one decade. The dashed line represents $\Delta N \propto \langle N \rangle^1$ to guide the eye, which is the prediction from the linear theory by Ramaswamy et al. [8]. Inset: the orientational correlation function $G_2(r) = \langle \cos 2(\theta_i - \theta_j) \rangle \rangle$, where i, j run over pairs of particles separated by a distance r, and θ_i, θ_j denote orientations of particles. For $N_{\text{total}} \geq 2500$ (upper two data), $G_2(r)$ exhibits algebraic decay, characteristic of quasi-long-range order, again only over one decade. Figures modified and reproduced from [19]. Reprinted with permission from AAAS(Readers may view, browse, and/or download material for temporary copying purposes only, provided these uses are for noncommercial personal purposes. Except as provided by law, this material may not be further reproduced, distributed, transmitted, modified, adapted, performed, displayed, published, or sold in whole or in part, without prior written permission from the publisher.)

Vibrated polar disks

The Olivier Dauchot group devised a monolayer system of polar disks vibrated on a horizontal plate [29] as shown in Fig. 2.13. Each particle has a symmetrical circular shape for avoiding nematic interactions usually present in granular systems, but its center of mass is displaced from its geometrical center. Hence when vibrated vertically, it moves persistently according to its polarity. Particles undergo polar alignment during collisions due to self-propulsion and hard-core repulsion. At high density, they observed ordered motion at the center of the system showing 'GNF' with the exponent $\alpha \sim 0.725$ (Fig. 2.13d), but later this 'GNF' were found not to be the genuine GNF predicted in the TTR phases.

The most significant problem they suffered was that they could not disentangle three possible number fluctuations: those from the boundary effects, those arising from the proximity of the transition, and the genuine GNF. Although their measurement was done in the most ordered conditions they could attain, their later study [34] proved that, unfortunately, their experiments were done right in the transitional region, so the 'GNF' obtained in their experiments are not the genuine GNF predicted

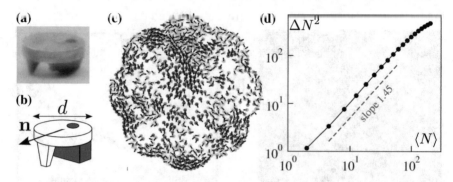

Fig. 2.13 **a** Photo of the fabricated polar disk. **b** Schematic of the polar disk. Its center of mass is displaced from its geometrical center, which leads to persistent motion under vibration. **c** Snapshot of the experiment. Overlaid colors indicate the directions of particles. Boundary effects and clustering are hardly removable. **d** 'GNF' obtained in this experiment. Figures (a) and (b) reproduced from [34], (c) from [29]. Figure (d) is modified from [29]

in the TTR phases. Furthermore, again, this experiment suffered from boundary effects and small system size. In this study, the diameter of the particles is $d_0 = 4$ mm and the diameter of the shaken plate is $\sim$160 mm. The particles are largely affected by the boundary and they often accumulate along the boundary. To exclude such boundary effects, they elaborated a flower-shaped boundary that 'reinjects' particles accumulated along the boundary into the bulk. Although they elaborated such nice conditions, such a boundary can also reinject clusters of particles into the bulk, which is unfavorable to realize the homogeneous TTR phases. They also need to decrease the size of their observation region of interest (ROI) for analysis in order to surely extract their bulk behavior. The size of ROI that they could reliably use for extracting bulk statistics was up to the diameter $20d_0 = 80$ mm. At the highest density of their experiments, they used 890 particles but only $\sim$160 particles were inside the ROI on average. Hence, the system size is not so large.

Although they could not bring their system to function deep in the ordered phase experimentally, their numerical simulation with periodic boundary conditions could exhibit a true long-range ordered phase with the genuine GNF, which is a TTR phase.

Bullets shaken in a sea of spherical beads

Kumar, Ramaswamy et al. conducted a nice experiment combining the knowledges on granular active matter experiments. They prepared millimeter-sized tapered rods like bullets and put them in a monolayer sea of spherical beads on a flower-shaped plate (Fig. 2.14) [30]. Under vertical vibrations, the bullets move inside the sea of beads, eventually align to each other, and exhibit polar order along the boundary at sufficiently high density. Their interactions are mediated by the flow of beads in between them.

Of course, this granular experiment again has a limited system size. The number of self-propelling bullets are $\sim$ 300, which is even smaller than other granular experiments. However, they also conducted surprisingly faithful three-dimensional

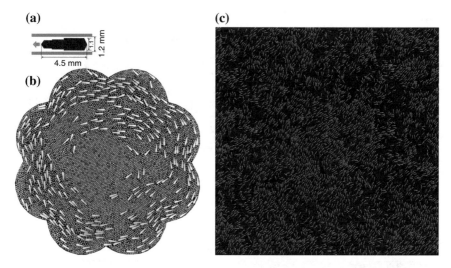

Fig. 2.14 **a** Schematic of a tapered bullet-like rod. Such rods are dispersed in the sea of spherical beads between two horizontal walls. **b** Collective flocking state obtained in this experiment. Due to the boundary, the obtained order is polar order along the boundary. **c** Homogeneous long-range ordered phase obtained in a faithful numerical simulation. The genuine GNF are observed only numerically. Figures reproduced from [30]

granular simulations with periodic boundary conditions using experimentally extracted parameters. These simulations clearly demonstrated the existence of GNF and the true long-range order in the ordered phase, hence we can conclude that this experimental system can *ideally* exhibit a TTR phase (Fig. 2.14c). However, accessing that phase experimentally is impossible due to the limited system size and the boundary effects.

Swarming bacteria on agar plates

Zhang, Swinney et al. reported 'GNF' in bacteria *Bacillus subtilis* swarming on an agar plate as shown in Fig. 2.15a [20]. Bacteria are micrometer-sized, so bacterial experiments are less affected by boundaries and easier to observe bulk statistics than granular experiments. However, the regimes they observed were not globally-ordered (see Fig. 2.15a, c). What they observed as 'GNF' is number fluctuations originating from the existence of finite-size ordered clusters moving around in different directions. The existence of clusters trivially increase the density fluctuations because the observed particle number fluctuates quite a lot depending on whether there are any clusters inside the field of view at that time or not (see Fig. 2.15c). Their orientation correlation and the velocity correlation were calculated inside the ordered clusters. Even inside such ordered clusters, the analyzed orientation/velocity correlations decay so rapidly at the length scale as small as 8 μm (Fig. 2.15b), which is smaller than 2 bacterial body lengths. They also depend on the cluster size (Fig. 2.15b). Therefore, the system is actually short-ranged, and cannot be a TTR phase.

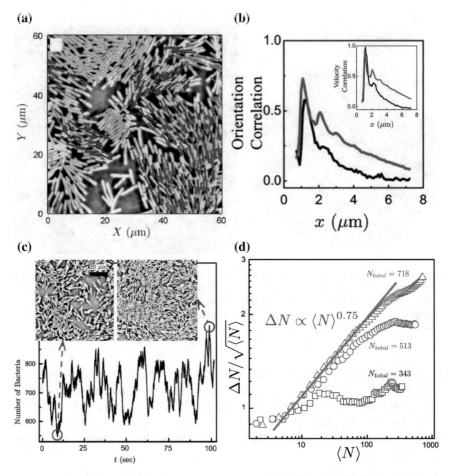

Fig. 2.15 **a** Snapshot of the experiment with bacterial velocity vectors overlaid. Colors of arrows indicate different 'clusters', which clearly demonstrates both the existence of clusters and the absence of long-range order. **b** Correlation of bacterial orientations measured inside clusters. Inset: Correlation of bacterial velocity measured inside clusters. Black: inside a cluster with 343 particles. Red: inside a cluster with 718 particles. Both correlations decay so rapidly at the length scale ∼8 μm, which demonstrates the short-range order of the system. **c** Time series of the number of observed bacteria. Its fluctuations are caused by clusters. **d** 'GNF' obtained in this experiment, originating from clustering without any long-range order. Green line: $\Delta N \propto \langle N \rangle^{0.75}$. Figures modified and reproduced from [20]

Gliding myxobacteria

Myxobacteira *Myxococcus xanthous* glide on an agar surface, and exhibit pattern formation and social behavior such as formations of fruiting bodies. Peruani et al. investigated collective motion of a mutant strain of *M. xanthus* that does not exhibit

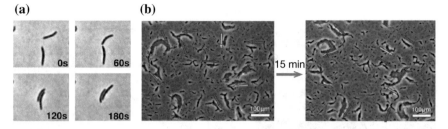

Fig. 2.16 **a** Myxobacteria *Myxococcus xanthous* exhibit nematic alignment upon collisions. However, because of strict two-dimensionality, they cannot cross or overlap each other. In this sense, their interactions are qualitatively different from the Vicsek-style self-propelled rods. **b** Collective phase of *M. xanthous*. There are many clusters moving in different directions. Arrows indicate the moving directions of the clusters. The average speed of bacteria is 3.10±0.35 µm/min. Figures modified and reproduced from [35]

directional reversals of their motility [35]. Such simple motility without reversal is suitable for testing theoretical predictions, especially that of self-propelled rods.

These myxobacteria show phase separation into dense ordered clusters moving around in different directions that do not order globally (Fig. 2.16). They do exhibit 'GNF', which is a trivial consequence of the phase separation and clustering. Hence, as originally claimed in their paper, the reported behavior was qualitatively different from that of the Vicsek-style models, but they succeeded to reproduce some of the behavior by a simulation on explicit rods with excluded volume.

Mesoscale turbulence of bacteria

In papers by Wensink et al. [32, 36], they investigated turbulent phases of dense suspensions of *Bacillus subtilis*. They conducted both experiments and simulations, and found 'GNF' only in their simulations [36]. In their case, number fluctuations were inaccessible in their experiments. They simulated $\sim 10^4$ self-propelled rods composed of connected chains with exclusive volume in a two-dimensional space, which are different from the Vicsek-style self-propelled rods model. This model exhibit many phases depending on the aspect ratio and the density of rods, and they obtained 'GNF' in swarming phases with clusters and in incoherent disordered phases *without* long-range order (Fig. 2.17). All the 'GNF' results were limited to numerical simulations.

Fibroblast cells

Duclos et al. investigated collective behavior of a monolayer sheet of fibroblast cells [31]. The fibroblast cells have elongated spindle-shapes, align nematically, and exhibit back-and-forth motion with no preferred directions. The obtained phase at high density is a domain phase with many topological defects (Fig. 2.18a).[7] The fibroblast cells do form locally aligned domains but do not exhibit long-range order.

---

[7]Such topological phases are often observed in two-dimensional cultures of mammalian cells with high aspect ratio such as neural progenitor cells [37] and epithelial cells [38].

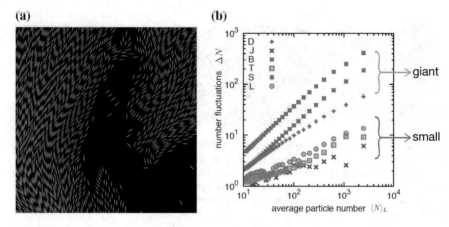

Fig. 2.17 **a** Snapshot of the numerically obtained 'swarming phase'. Large clusters exist and no long-range order is observed. **b** 'GNF' reported only for such swarming phases (S: swarming, B: bionematic) and the incoherent disordered phase (D: disordered), both of which do not have long-range order. Other numerically obtained phases only exhibit smaller fluctuations with $\alpha < 0.5$ due to excessively high density and excluded volume. Figures modified and reproduced from [32]

The correlation function of their orientation field was quite nicely fitted by an exponential function, which clearly demonstrates that there exists a finite correlation length ξ and they only have *short-range* order (Fig. 2.18b).

Furthermore, there is another puzzling result with this experiment. Duclos et al. observed 'GNF' even at low density with very small nematic order before confluence. In Fig. 2.18, they obtained 'GNF' even at the lowest density they observe, which is 10 times more dilute than that of confluence. Therefore, it makes even difficult to extract what is the dominant origin of these 'GNF'.

Although Duclos et al. reported 'GNF' in such a domain phase with short-range order, again this is not what we can compare with the series of studies on the Vicsek-style models and corresponding hydrodynamic theories.

Actomyosin motility assay

Systems of biofilaments driven by molecular motors have been devised and it turned out to exhibit collective motion at high density of filaments [27, 39–41]. An experimental technique called 'motility assay' is utilized for studies on collective motion thanks to its controllability.

Motility assays were originally used to investigate behavior of molecular motors at the single molecule levels. In motility assays, molecular motors such as myosins, dyneins, and kinesins are attached and fixed on a glass substrate. Then the cytoskeletal filaments such as actin filaments and microtubules are spread onto the carpet of the molecular motors, and accordingly those filaments are driven by the molecular motors and move around. In biophysical studies on molecular motors, properties of molecular motors such as their forces, fluctuations, and step-wise motion can be extracted from motility assay experiments at low density of filaments.

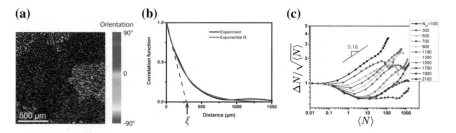

Fig. 2.18 **a** Snapshot of the domain phase obtained in a monolayer of fibroblast cells. Colors are overlaid according to the orientations of cells. **b** Correlation function of the orientation field. It is very nicely fitted by an exponential function, which suggests the existence of a characteristic length scale and short-range order. **c** 'GNF' are obtained even in the disordered phase at low density as well as in the domain phase after confluence. The total number of particles in the field of view $N_{tot} \simeq 1000$ corresponds to the confluence. Figures reproduced from [31]

In the studies of collective motion, the density of filaments is some order of magnitude larger than single molecule studies. As the density increases, we can observe transitions from disordered phases with almost independent motion of filaments to coordinated ordered motion [27, 39–41]. Because motility assay systems are composed of nano-/micro-scale filaments, the number of filaments in the whole system can be huge. Furthermore, interactions and motility can be tuned by changing lengths of filaments and density of driving motor proteins, adding polymers into solutions, and controlling the height uniformity of motor proteins attached on the substrate. Due to such controllability and large system size, motility assay systems are now becoming very important model systems in active matter physics.

In the motility assay system with actin filaments and myosin motors—actomyosin systems—(Fig. 2.19a), 'GNF' were reported by the Andreas Bausch group [27]. However, they could not achieve homogeneous ordered phases as the TTR phase. The system only exhibits swarming clusters whose directions of motion change continuously and a banding phase similar to the Vicsek wave (Fig. 2.19c). The 'GNF' in this system were measured when the global flow is not straight and the system is disordered on large-scales as shown in Fig. 2.19b,d. They do obtain homogeneously ordered phase only at the beginning of experiments as a transient state, but this ordered state can only show normal fluctuations with the exponent $\alpha \sim 0.5$ (see the red data points and the caption in Fig. S5B of the Supplementary Material of Schaller and Bausch [27]).

Quincke rollers

By utilizing an electrohydrodynamic phenomenon called Quincke rotation, the Denis Bartolo group realized self-propelled colloids rolling on a two-dimensional electrode placed horizontally [28]. They applied a vertical electric field to insulating spheres immersed in a conducting fluid between two horizontal two-dimensional electrodes. As shown in Fig. 2.20a, above a certain critical amplitude of the electric field, the charge distribution induced on the surfaces of the spheres gets unstable to

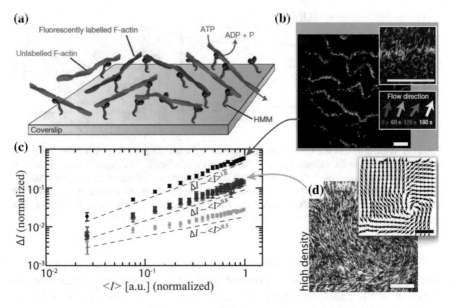

Fig. 2.19 **a** Schematic of the actomyosin motility assay system. **b** Number fluctuations are estimated by the fluorescent intensity I of images. **c** At banding states similar to the Vicsek wave, they naturally observe the exponent $\alpha \simeq 1$ due to clustering. In the dilute regime, $\alpha \simeq 0.5$ is obtained as expected. **d** In the collective phase at high density *without* long-range order, they observe 'GNF' with $\alpha \simeq 0.8$. Topological defects exist and there is no global order. Inset: velocity field. Scale bars: 50μm Figures (a) (c) and (b) (d) modified and reproduced from [39] and [27] respectively

infinitesimal fluctuations. The symmetry of charge distribution spontaneously gets broken, which results in an electrostatic torque driving particles' rotations. Such rotations are then transformed into translational motion on a two-dimensional electrode.

These rolling colloids interact with each other via both hydrodynamic and electrostatic interactions. These interactions effectively align the particles in polar configurations, and at high enough density they observed macroscopically ordered colloidal flocks moving in the same direction. The transition from the disordered phase to the ordered phase was clearly observed with a coexisting phase at the transitional region and such behavior is nicely explained by theoretical calculations on the interactions of particles (Fig. 2.20b–e).

However, the number fluctuations obtained in this system were considered to be only normal with the exponent $\alpha \sim 0.5$ (Fig. 2.20f). This behavior was also explained by their theoretical calculation that long-range hydrodynamic interaction had suppressed the density fluctuations, leading to normal fluctuations. Such an additional long-range interaction absent in the Vicsek-style models can hinder experimental observation of GNF in the TTR phases. Here we note that their recent careful analysis on the Quincke roller experiments has finally detected GNF [42] which was published 1.5 years after this thesis and our experiment [26] presented in Chap. 3.

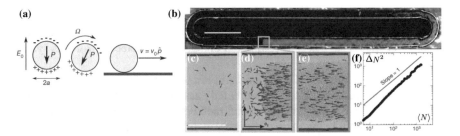

Fig. 2.20 **a** Schematics of the mechanism of Quincke rotation and subsequent self-propulsion on a plane. **b** Dark-field image of a flock of the Quincke rollers in a channel. Scale bar: 3 mm. **c** Isotropic disordered phase. **d** Banding phase. **e** Homogeneous ordered phase. Scale bar: 500mm. **f** Number fluctuations in the ordered phase like **e** are normal with $\alpha = 1$. Note that in this plot the variance ΔN^2 is shown instead of the standard deviation ΔN. Figures modified and reproduced from [28]

2.6.3 Summary of Experimental Approaches

As we have detailed above, there are many studies reporting 'GNF' due to the relatively easy experimental accessibility of number fluctuations. However, most of those 'GNF' were originating from clustering or boundary effects, which are not deeply rooted in the mathematical properties of symmetry broken states. Understanding the theories on collective motion and implementing theoretical requirements in experimental systems simultaneously have actually been quite difficult due to many pitfalls. This situation makes people unfamiliar with the theoretical background confused to think that the 'GNF', or sometimes even active matter physics, are trivial and not so interesting.

Vibrated grains are a relatively easy realization of active systems and often give us new insight [19, 29, 30]. However, they always suffer from the small system size and the boundary effects due to the millimeter-size of the particles. Although corresponding faithful simulations sometimes clearly demonstrate that those experimental systems can belong to the Vicsek universality class [29, 30], the TTR phase is quite difficult to achieve experimentally.

Bacterial systems do not so much suffer from the small system size or the boundary effects, due to their micrometer-size. However, it is rather difficult to control their interactions and the existing studies could only achieve clustering phase [20, 35] or turbulent phase [32] without global long-range order.

Motility assays are rather controllable experiments, although they require sophisticated experimental techniques. However, for the moment, in all the reported motility assay systems, filaments at high density often form clusters/bands [27, 39] or cannot move straight persistently [40]. Therefore they do not possess long-range order. Furthermore, their interactions are still unclear. For example, there is some debate on whether hydrodynamic interactions play some role in such motility assay systems or not.

Quincke rollers were a nice system very close to exhibit a TTR phase, but unfortunately their long-range hydrodynamic interactions were considered to tame giant density fluctuations at that time [28]. Later on, GNF were detected with this system [42].

In summary, it is difficult but important to experimentally realize homogeneous but highly-fluctuating orientationally-ordered phase with long-range order. The existence of long-range order should be confirmed before moving on to the discussion on the genuine GNF in the sense of the Vicsek universality class.

So, then, what kind of experimental system is required for realizing the Toner-Tu-Ramaswamy phenomena that belong to the Vicsek universality class? We will answer this question in the next chapter.

References

1. Chaté H, Ginelli F, Grégoire G, Raynaud F (2008) Collective motion of self-propelled particles interacting without cohesion. Phys Rev E 77(4):046113
2. Vicsek T, Czirók A, Ben-Jacob E, Cohen I, Shochet O (1995) Novel type of phase transition in a system of self-driven particles. Phys Rev Lett 75(6):1226
3. Grégoire G, Chaté H (2004) Onset of collective and cohesive motion. Phys Rev Lett 92(2):025702
4. Toner J, Tu Y (1995) Long-range order in a two-dimensional dynamical XY model: how birds fly together. Phys Rev Lett 75(23):4326–4329
5. Toner J, Tu Y (1998) Flocks, herds, and schools: a quantitative theory of flocking. Phys Rev E 58(4):4828–4858
6. Toner J (2012) Reanalysis of the hydrodynamic theory of fluid, polar-ordered flocks. Phys Rev E 86(3):031918
7. Toner J, Tu Y, Ramaswamy S (2005) Hydrodynamics and phases of flocks. Ann Phys 318:170–244
8. Ramaswamy S, Simha RA, Toner J (2003) Active nematics on a substrate: giant number fluctuations and long-time tails. Europhys Lett 62(2):196–202
9. Ginelli F (2016) The physics of the vicsek model. Eur Phys J: Special Topics 225:2099–2117
10. Tu Y, Toner J, Ulm M (1998) Sound waves and the absence of galilean invariance in flocks. Phys Rev Lett 80(21):4819–4822
11. Peshkov A, Aranson IS, Bertin E, Chaté H, Ginelli F (2012) Nonlinear field equations for aligning self-propelled rods. Phys Rev Lett 109(26):268701
12. Bertin E, Chaté H, Ginelli F, Mishra S, Peshkov A, Ramaswamy S (2013) Mesoscopic theory for fluctuating active nematics. New J Phys 15(8):085032
13. Peshkov A, Bertin E, Ginelli F, Chaté H (2014) Boltzmann-Ginzburg-Landau approach for continuous descriptions of generic Vicsek-like models. Eur Phys J Special Topics 223:1315–1344
14. Mermin ND, Wagner H (1966) Absence of ferromagnetism or antiferromagnetism in one- or two-dimensional isotropic Heisenberg models. Phys Rev Lett 17(22):1133
15. Kosterlitz JM, Thouless DJ (1973) Ordering, metastability and phase transitions in two-dimensional systems. J Phys C: Solid State Phys 6:1181–1203
16. Chaté H, Ginelli F, Montagne R (2006) Simple model for active nematics: quasi-long-range order and giant fluctuations. Phys Rev Lett 96(18):180602
17. Ngo S, Peshkov A, Aranson IS, Bertin E, Ginelli F, Chaté H (2014) Large-scale chaos and fluctuations in active nematics. Phys Rev Lett 113(3):038302

18. Ginelli F, Peruani F, Bär M, Chaté H (2010) Large-scale collective properties of self-propelled rods. Phys Rev Lett 104(18):184502
19. Narayan V, Ramaswamy S, Menon N (2007) Long-lived giant number fluctuations in a swarming granular nematic. Science (New York, N.Y.), Vol 317, p 105
20. Zhang HP, Be'er A, Florin E-L, Swinney HL (2010) Collective motion and density fluctuations in bacterial colonies. Proc Natl Acad Sci USA, Vol 107, No 31, pp 13626–13630
21. Mishra S, Puri S, Ramaswamy S (2014) Aspects of the density field in an active nematic. Philos Trans Royal Soc A 372(2029):20130364
22. Mishra S, Simha RA, Ramaswamy S (2010) A dynamic renormalization group study of active nematics. J Stat Mech: Theory Exp 2010(02):P02003
23. Marchetti MC, Joanny JF, Ramaswamy S, Liverpool TB, Prost J, Rao M, Simha RA (2013) Hydrodynamics of soft active matter. Rev Modern Phys 85(3):1143–1189
24. Solon AP, Tailleur J (2013) Revisiting the flocking transition using active Spins. Phys Rev Lett 111(7):078101
25. Solon AP, Chaté H, Tailleur J (2015) From phase to microphase separation in flocking models: the essential role of nonequilibrium fluctuations. Phys Rev Lett 114(6):068101
26. Nishiguchi D, Nagai KH, Chaté H, Sano M (2017) Long-range nematic order and anomalous fluctuations in suspensions of swimming filamentous bacteria. Phys Rev E 95(2):020601(R)
27. Schaller V, Bausch AR (2013) Topological defects and density fluctuations in collectively moving systems. Proc Natl Acad Sci USA 110(12):4488–4493
28. Bricard A, Caussin J-B, Desreumaux N, Dauchot O, Bartolo D (2013) Emergence of macroscopic directed motion in populations of motile colloids. Nature 503(7474):95–98
29. Deseigne J, Dauchot O, Chaté H (2010) Collective motion of vibrated polar disks. Phys Rev Lett 105(9):098001
30. Kumar N, Soni H, Ramaswamy S, Sood AK (2014) Flocking at a distance in active granular matter. Nature Commun 5:4688
31. Duclos G, Garcia S, Yevick HG, Silberzan P (2014) Perfect nematic order in confined monolayers of spindle-shaped cells. Soft Matter 10(14):2346–2353
32. Wensink HH, Dunkel J, Heidenreich S, Drescher K, Goldstein RE, Löwen H, Yeomans JM (2012) Meso-scale turbulence in living fluids. Proc Natl Acad Sci USA 109(36):14308–14313
33. Chaté H, Ginelli F, Grégoire G, Peruani F, Raynaud F (2008) Modeling collective motion: variations on the Vicsek model. Eur Phys J B 64:451–456
34. Weber CA, Hanke T, Deseigne J, Léonard S, Dauchot O, Frey E, Chaté H (2013) Long-range ordering of vibrated polar disks. Phys Rev Lett 110(20):208001
35. Peruani F, Starruß J, Jakovljevic V, Søgaard-Andersen L, Deutsch A, Bär M (2012) Collective motion and nonequilibrium cluster formation in colonies of gliding bacteria. Phys Rev Lett 108(9):098102
36. Wensink HH, Löwen H (2012) Emergent states in dense systems of active rods: from swarming to turbulence. J Phys: Condensed Matter Condensed Matter 24(46):464130
37. Kawaguchi K, Kageyama R, Sano M (2017) Topological defects control collective dynamics in neural progenitor cell cultures. Nature 545:327–331
38. Saw TB, Doostmohammadi A, Nier V, Kocgozlu L, Thampi S, Toyama Y, Marcq P, Lim CT, Yeomans JM, Ladoux B (2017) Topological defects in epithelia govern cell death and extrusion. Nature 544(7649):212–216
39. Schaller V, Weber C, Semmrich C, Frey E, Bausch AR (2010) Polar patterns of driven filaments. Nature 467(7311):73–77
40. Sumino Y, Nagai KH, Shitaka Y, Tanaka D, Yoshikawa K, Chaté H, Oiwa K (2012) Large-scale vortex lattice emerging from collectively moving microtubules. Nature 483:448–452
41. Suzuki R, Weber CA, Frey E, Bausch AR (2015) Polar pattern formation in driven filament systems requires non-binary particle collisions. Nature Phys 11:839–844
42. Geyer D, Morin A, Bartolo D (2018) Sounds and hydrodynamics of polar active fluids. Nature Mater 17(9):789–793

Chapter 3
Collective Motion of Filamentous Bacteria

Abstract This chapter describes our experimental study on collective motion of long, filamentous, non-tumbling bacteria swimming in a thin fluid layer. The confinement in the quasi-two-dimensional plane and the high aspect ratio of the cells induce weak nematic alignment upon collision due to the weak excluded volume interactions, which, for sufficiently high density of cells, gives rise to global nematic order. This homogeneous but highly-fluctuating phase, observed on the largest experimentally-accessible scale of millimeters, exhibits the properties predicted by the standard flocking models, especially the Vicsek-style self-propelled rods: true long-range nematic order and non-trivial giant number fluctuations. Thus, our experimental system is recognized as the first unambiguous example that falls into the Vicsek universality class. Our results suggest necessary conditions for the Vicsek universality class in comparison with other experimental studies. This chapter is based on our publication [Nishiguchi et al., Physical Review E 95, 020601(R) (2017)] but also includes further detailed analysis of correlations and collisions.

Keywords Elongated bacteria · Self-propelled rods · Giant number fluctuations · Long-range order · Vicsek universality class

3.1 Introduction

As we have detailed in Chap. 2, standard models on collective motion—the Vicsek style models [1–6] and corresponding hydrodynamic theories [7–12]—have revealed their universal properties, which are often distinctively different from those of equilibrium orientationally-ordered phases. In particular, the spontaneous breaking of continuous rotational symmetry and the crucial coupling between the orientation and the density field generate anomalously-large giant number fluctuations (GNF) along with algebraic correlations of orientation and density in their long-range ordered phases. Such phases are now called 'the Toner-Tu-Ramaswamy phases (the TTR phases)'.

Experimental attempts to find the TTR phases have failed because of many experimental pitfalls and overlooked theoretical requirements as we have detailed in

© Springer Nature Singapore Pte Ltd. 2020   45
D. Nishiguchi, *Order and Fluctuations in Collective Dynamics*
of Swimming Bacteria, Springer Theses, https://doi.org/10.1007/978-981-32-9998-6_3

Sect. 2.6. Therefore, although there exist a huge number of theoretical and numerical studies on collective motion which are mostly based on the Vicsek-style models, their experimental basis has been lacking. The lack of experimental evidence for such universality has also hindered the understanding of necessary/sufficient conditions for realizing the Toner-Tu-Ramaswamy phenomena in real systems.

Here we study the collective dynamics of elongated microswimmers in a very thin fluid layer between two walls by devising long, filamentous, non-tumbling bacteria. The strong confinement and the high aspect ratio of cells induce weak nematic alignment upon collision, which, for sufficiently high density of cells, gives rise to global nematic order. This homogeneous but highly-fluctuating phase, observed on the largest experimentally-accessible scale of millimeters, exhibits the properties predicted by standard flocking models, especially the Vicsek-style self-propelled rods (polar particles with nematic interactions) [5]: true long-range nematic order and non-trivial GNF. Therefore, our experimental system has turned out to fall into the Vicsek universality class, and gives the first unambiguous example exhibiting the Toner-Tu-Ramaswamy phase.

3.2   Strategy: Use of Filamentous Cells in Confinement

The collective behavior of bacteria is a vast topic of research with obvious and crucial biological interest. Bacteria have also been widely used by physicists as attractive active matter systems because their small size enables large system size experiments far away from boundaries. Both crawling/sliding and swimming bacteria have been used, but so far, no very long-range ordering/collective motion has been observed as we have detailed in Sect. 2.6. Sliding myxobacteria, for example, align, collide and form very dense ordered clusters, but these clusters are of limited size, being easily destroyed upon collision and moving in various directions [13]. *Bacillus subtilis* swimming/swarming on agar surfaces form loose ordered clusters with anomalous density fluctuations, again of limited size and moving in various directions [14]. Dense suspensions of swimming bacteria typically give rise to 'bacterial turbulence' [14–19], i.e. a chaotic regime with a dominant length scale of about 10–20 cell lengths. We will detail such turbulent states, 'active turbulence', in Chap. 4.

Two factors are often invoked to explain this situation:

- Long-range hydrodynamic interactions are theoretically known to destabilize ordered states [20–23].
- The aspect ratio of cells is too small to lead to strong alignment upon collision [17].

As for the first factor, the hydrodynamic flow field created by a single bacterium can be nicely approximated by flow created by a force dipole [24]. Bacteria swim by pushing fluid behind them, and accordingly their heads also push fluid in front of them. Therefore, a single bacterium exerts two forces in opposite directions, and it can be represented by a force-dipole directing outward. In this sense, bacteria are so-called 'pusher-type microswimmers'. Dipolar flow created by bacteria are schematically shown in Fig. 3.1. If two bacteria are aligned in a parallel configutation, they push each other by their dipolar flow and such a configuration gets unstable, leading to bended configurations as shown in the right side of Fig. 3.1. Hence, when hydrodynamics can play significant role in bacterial dynamics, globally ordered states cannot develop and what we observe is turbulent states without long-range order [15–19]. These explanations were also confirmed by numerical simulations on Stokesian dynamics [20–23, 25].

As for the second factor, even without hydrodynamics, it is still difficult to achieve long-range ordered states. As we have seen already, bacterial experiments on agar plates, in which bacteria are gliding/crawling and hydrodynamics does not apparently play significant roles, also cannot exhibit long-range order [13, 14]. In numerical studies [17, 26] on self-propelled rods with exclusive volumes and without hydro-dynamics, they explored how aspect ratios affect macroscopic behavior. They found that, even without hydrodynamics, self-propelled rods with aspect ratio $\lesssim 14$ cannot develop global order and result in turbulent states at high density. However, they achieved to observe laning states in which counter-going lanes coexist with aspect ratios $\gtrsim 14$. Although these states are different from the ordered phase observed in the Vicsek-style models and corresponding field equations,[1] we can learn that suf-ficiently high aspect ratios are required for attaining global order. Commonly used bacteria cannot reach such high aspect ratios. Typical aspect ratios of bacterial bod-ies are ~ 3 for *E. coli* and ~ 6 for *B. subtilis*, which nicely correspond to turbulent regimes in numerical studies [17, 26].

Fig. 3.1 Flow field created by bacteria destabilizes aligned configurations of bacteria. This instability prevents dense bacterial suspensions from exhibiting global order, resulting in turbulent states

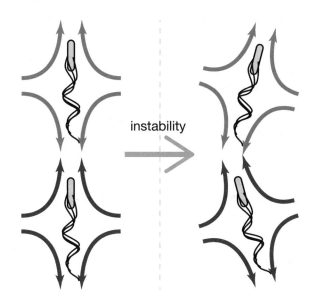

instability

---

[1] This is because particles cannot overlap in this model due to excluded volume but can overlap in the Vicsek-style point-like particles models.

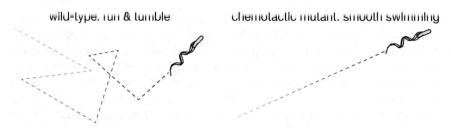

Fig. 3.2 Schematic figures of run and tumble motion of wild-type bacteria and smooth swimming of chemotactic mutant bacteria. Usual bacteria which are wild-type for chemotaxis exhibit so-called run & tumble motion. At tumbling phases, they change the directions of motion in order to climb up/down chemical gradients. On the other hand, chemotactic mutant bacteria like the strain PR49479 used in our study do not exhibit tumbling motion and just swim smoothly

To prevent these two pitfalls, we devised a system of filamentous cells of *Escherichia coli* bacteria [27, 28] confined between two solid walls with a small, micrometer-sized gap. Confinement suppresses fluid flow created by bacteria and thus prevents instability of the ordered states. By making filamentous cells of *E. coli*, we achieved higher aspect ratios.

3.3   Experimental Procedure and Setup

3.3.1   Preparation of Non-tumbling Filamentous Cells

The filamentous bacteria were obtained by incubating usual *E. coli* cells under the influence of the antibiotic 'cephalexin' (20 µg/ml), which allows cell growth but inhibits cell division.

To avoid complicated dynamics and possible artifacts originating from chemotactic behavior, we used a non-tumbling chemotactic mutant strain of *Escherichia coli* (RP4979, ΔcheY, [29]) that had been transformed to express a yellow fluorescent protein (plasmid: pZA3R-YFP) (Fig. 3.2). The plasmid used here contains a gene for resistance against an antibiotic chloramphenicol. The strain RP4979, unlike a wild type stain RP437, lacks CheY protein responsible for flagella rotational switch. This CheY deleted mutant exclusively rotates flagella in counterclockwise (CCW) direction and swim persistently without tumbling.

The bacteria taken from a frozen stock were grown overnight for 16 h in Luria Broth (LB) with a selective antibiotic (chloramphenicol 33 µg/ml) shaken at 200 rpm at 30 °C. Then this culture was diluted 100-fold in 10ml of Tryptone Broth (TB, 1 wt% tryptone and 0.5 wt% NaCl) with the selective antibiotic (chloramphenicol 33 µg/ml) and incubated in a 125 ml flask shaken at 200 rpm at 30 °C. We waited for bacteria to reach the exponential growth phase. After 2 h, the antibiotic cephalexin was added at the final concentration 20 µg/ml and we continued growing bacteria

(a)   **(b)**

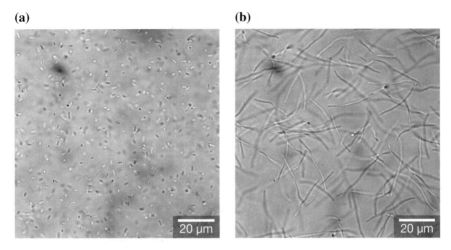

Fig. 3.3 Bright field images of **a** normal cells and **b** filamentous cells. Contrast and brightness are adjusted for visibility. Lengths of normal cells are ~2–3 μm. After adding the antibiotic 'cephalexin', cells start to elongate and their lengths can reach more than 20–100 μm. Diameters of bacteria are smaller than 1 μm, and around 0.8 μm

for another 3 h to obtain filamentous cells (Fig. 3.3). Thus we obtained sufficiently dense suspensions of filamentous cells with both high motility and moderate body lengths.

The filamentous cells have flagella all around their bodies at the same density as usual bacteria, and are still able to swim [27, 28]. Although lengths of cells can be controlled by varying the duration of incubation after adding the antibiotic, their swimming speed gradually decreases with their lengths [28, 30]. We chose moderate body lengths of $\sim 19 \pm 5$ μm ($\pm$: standard deviation) to obtain cells with both sufficient nematic interactions and sufficient swimming speed (Fig. 3.4). Thus the filamentous cells above have aspect ratios of ~25.

To concentrate the obtained suspension, 1 ml of the suspension was mildly passed through a membrane filter with 0.22-μm pores (Merck Millipore, Isopore GTBP01300) and we retrieved the concentrated suspension from the membrane.[2]

3.3.2 Observation Devices

The suspension of filamentous cells, after concentration, was placed on a coverslip (MATSUNAMI, thickness 0.12–0.17 mm) and then sealed with a polydimethylsiloxane (PDMS) plate without any spacers to make the gap as small as possible (Fig. 3.5).

---

[2]We used a membrane filter, because concentrating suspensions by centrifugation does not work well. Centrifugation often damages such long filamentous bacteria, leading bacteria to die or to be almost non-motile.

Fig. 3.4 Distribution of bacterial lengths before (blue, left) and after (yellow, right) one typical experiment. The average lengths were 18.86 ± 0.21 µm and 18.83 ± 0.21 µm (mean $\pm$ standard error) respectively. Thus there is no detectable change of bacterial lengths during the experiment. Lengths of bacteria were measured by fitting polygonal lines with ImageJ

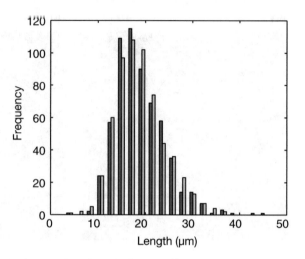

The PDMS plate was patterned with some wells which the excess fluid can escape into and can work as bacterial reservoirs (Fig. 3.6). The well patterns on the PDMS plate were fabricated by a standard soft-lithography technique using a photoresist SU-8. We thus achieved a gap of about $\sim$2 µm. We used PDMS because it transmits oxygen required to sustain higher motilities of bacteria. Thanks to the permeability of PDMS to oxygen, typical experiments could be run for about 30 min without discernible changes in the behavior of the cells. Prior to putting the suspension onto the coverslip, the coverslip and the PDMS plate were soaked in 1 wt% bovine serum albumin (BSA) solution for more than 1 h in order to prevent the bacteria from sticking on the surfaces. To reduce the gap width, we slightly pressed the PDMS plate. Such strong confinement contributed to suppressing the destabilizing fluid flow created by bacteria due to no-slip boundary conditions on the walls.

The confinement also helped preventing bacterial circular motion near solid walls [30] as depicted in Fig. 3.7. Because bacteria swim by rotating their flagella and their chiral symmetry is broken, bacteria exhibit circular swimming trajectories near a single wall due to the hydrodynamic coupling between their chiral propulsion mechanism and the solid wall. However, when bacteria are sandwiched between two solid walls with a sufficiently small gap width, the hydrodynamic effects by two walls compensate each other [31], enabling our bacteria to swim straight with some rotational diffusion over the largest distances (millimeters) considered below. Note that the gap width required for straight motion is larger for longer bacteria, so the use of the filamentous cells made it much easier to design our experimental setup.

After waiting for initial fluid flow—triggered when introducing the suspension—to be suppressed, we captured movies by a CMOS camera (Baumer HXG40, 2048 $\times$ 2048 pixels, 12 bit) at 5 Hz through an inverted fluorescent microscope (Leica DMi8 with Adaptive Focus Control) with an objective lens (HC PL FLUOTAR, 10$\times$, NA = 0.30). The area of the field of view was 1.12×1.12 mm^2, a size limited mainly by our will to be able to distinguish individual cells on the recorded images for

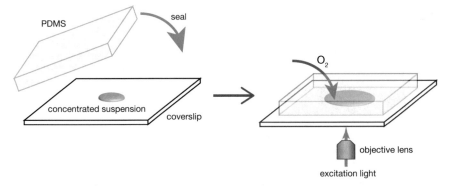

Fig. 3.5 Schematic figure of the experimental setup. Concentrated suspension of filamentous bacteria was placed on a coverslip and then a PDMS plate was put onto the suspension. Thanks to the permeability of PDMS to oxygen, sufficiently high motility of cells was kept during experiments. Well patterns on PDMS were not depicted for clarity

Fig. 3.6 Schematic figure of the well patterns made on PDMS. White circles are wells with the depths of ~10 μm. Wells are made on the PDMS plate to work as bacterial reservoirs that diminish boundary effects. Observation of ordered states is done in between these well patterns

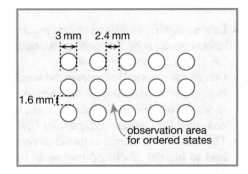

insuring reliability of our analysis, especially for analysis on GNF. The microscope was equipped with an adaptive autofocusing system to reduce unfavorable intensity fluctuations. The duration of the analyzed movies was 400 s (2000 frames) and there was no detectable change in bacterial lengths during the experiments as seen in Fig. 3.4. Hence this fact of no cell elongation together with inhibition of cell divisions insures that there is neither gradual increase of the total number of bacteria nor the area fraction, although in other usual bacterial experiments the number of bacteria doubles in approximately 20 min in the best conditions.

Here we note the importance of the choice of the magnification above. Because we are trying to extract large-scale long-time properties precisely from the fluorescent images, we need to probe as large an area as possible but the obtained images have to be reliable for estimating both nematic order and number fluctuations simultaneously.[3] As a result, we chose the above setup in accordance with the following constraints:

---

[3]Because we are trying to find GNF in the true long-range ordered states, both order parameter and number fluctuations have to be reliably quantified in the same images.

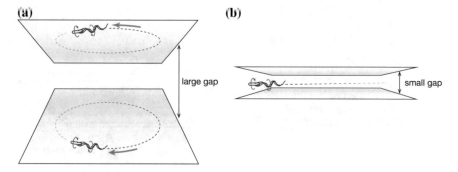

Fig. 3.7 Schematic figures of circular swimming motion near a single wall and straight swimming motion between two walls with a sufficiently small gap. Bacteria near a single wall exhibit circular trajectory due to the hydrodynamic coupling between their rotating flagella/bodies and the solid surface. However, when they are confined between two walls with a sufficiently small gap width, such effects are compensated and they swim straight

- Lower magnification observations require a stronger excitation light source in fluorescent microscopy. Such strong excitation light harms the bacteria and diminishes the duration of the observation.
- Otherwise, we need to increase the sensitivity (gain) of the camera, which results in stronger noise. However, we need to eliminate such noise as much as possible to see the reliable number fluctuations, because such noise can also contribute to unwanted fluctuations and *spurious* GNF.
- The use of lower magnification objectives also reduces the spatial resolution. This lead to blurred or dilated images of bacteria, and, consequently, our bacterial counting method using binarization (explained below) can become unreliable.

3.4 Analysis and Results

3.4.1 Overview of the System's Behavior

In experiments at low density of cells, or with a larger spacing ($\sim 10\,\mu m$) between the two surfaces, cells do not align enough to order on large scales (Fig. 3.8a, b). But at high concentration (average area fraction ~ 0.25), their collisions are so frequent that global nematic order emerges in spite of the weakness of alignment in terms of binary collisions. This ordered phase is strongly fluctuating but statistically homogeneous, without clusters. Bacteria then swim in opposite directions in approximately equal numbers (Fig. 3.8c, d). This nematic phase has turned out to exhibit long-range order and giant number fluctuations, which are consistent with the predictions on the Toner-Tu-Ramaswamy phase of the Vicsek-style self-propelled rods model.

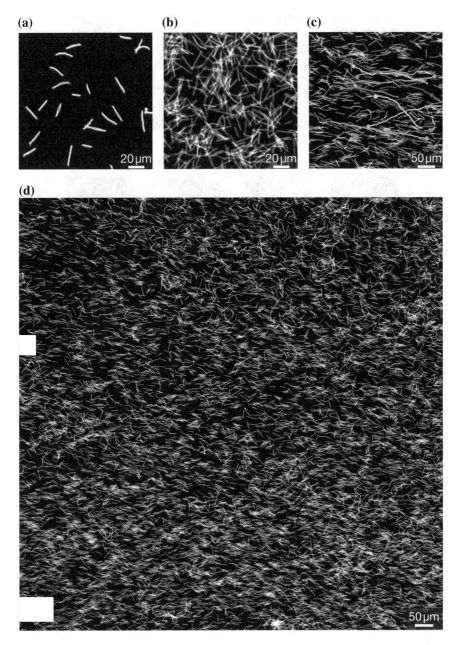

Fig. 3.8 Typical snapshots. **a** Zoom of the disordered phase at low density in a 2-μm thin experiment. **b** Zoom of the disordered phase at high density in a 10-μm thick experiment. **c** Zoom of the nematically-ordered phase at high density in a thin experiment with superimposed, manually tracked, 10-s trajectories of a few cells. **d** Full field of view in the same experiment as in (**c**)

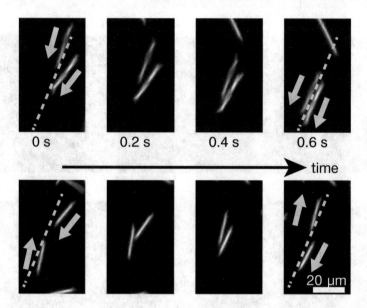

Fig. 3.9 Aligning collision events between two bacteria. Top: acute angle collision leading to alignment. Bottom: obtuse angle collision leading to anti-alignment. Dashed line: mean outgoing angle. The mean incoming angle is not shown for visibility, as it is only slightly different from the outgoing angle

3.4.2 Collision Statistics

To understand interactions of bacteria in this system, we first quantified their binary collisions. Our setup was thin enough to make it difficult for bacteria to cross each other without collisions. Some clear events of 'nematic alignment' upon collisions are shown in Fig. 3.9. We investigated binary collisions using movies taken at a relatively low density of bacteria and quantified interactions due to collisions. To decrease the number of multi-particle collisions, suspension of filamentous bacteria prepared in a way described above was diluted 3-fold in fresh Tryptone Broth with chloramphenicol. This diluted suspension was sandwiched between a coverslip and a PDMS plate with a small gap in the same manner as in the experiment for the global nematic phase, which we will describe later. We detected and tracked the center of mass of each bacterium from binarized images. We note that the movies for collision analysis here were captured through another microscope (Nikon ECLIPSE TE2000-U) with different objective lenses (Nikon Plan Fluor ELWD, 40×, NA = 0.60 for Fig. 3.9, and Nikon Plan Fluor, 10×, NA = 0.30 for Figs. 3.10 and 3.11).

Binary collisions were analyzed with binarized images. We defined the beginning of a collision as a merger event of two white objects in the binarized images that were isolated in the previous frame, and the end of the collision as a splitting event of that connected object. We inspected thousands of automatically detected merger events by eye and excluded multi-particle collisions. Thus we obtained 2204 collision events

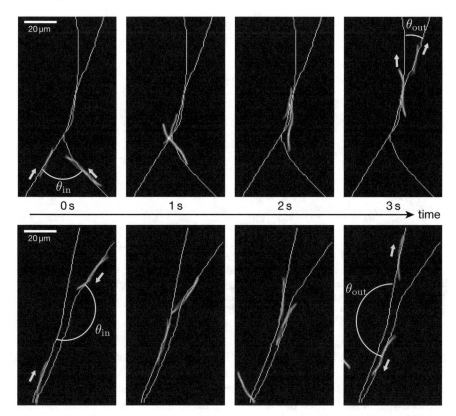

Fig. 3.10 Trajectories of colliding two bacteria and schematic definitions of the incoming angle θ_{in} and the outgoing angle θ_{out}. Trajectories of typical collision events with parallel alignment (top) and anti-parallel alignment (bottom) are superimposed on the experimental movies

with accurate tracking. To quantify binary collisions, we calculated an incoming angle θ_{in} and an outgoing angle θ_{out} for each collision event. θ_{in} and θ_{out} are defined as angles formed by instantaneous velocity vectors of two colliding bacteria just before and after the collision events respectively (Fig. 3.10). The velocity vectors are calculated from differences of positions in two successive frames separated by 0.2 s. Collision events with durations longer than 15 frames (3 s) are defined as complete polar alignment events ($\theta_{\text{out}} = 0°$). All the analyzed data are shown in Fig. 3.11. Red data points in Fig. 3.11 are mean values and standard deviations of θ_{out} calculated via binning θ_{in} at every 10°. We note that the number of observed events in acute angles is biased to be smaller than that in obtuse angles, because durations of collisions with acute angles are usually longer and the probability of ending up multi-particle collisions is higher.

As we can see from Fig. 3.11, on average, two bacteria incoming at some acute (obtuse) angle θ_{in} end up parallel (antiparallel). Overall, however, alignment is weak and many events do not result in such ideal nematic alignment with the outgoing angle

(a)   **(b)**

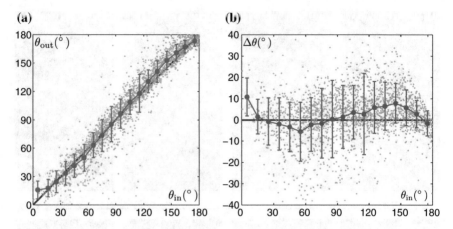

Fig. 3.11 Statistics on binary collisions show a nematic tendency. **a** Incoming angles θ_{in} versus outgoing angles θ_{out}. **b** Incoming angles θ_{in} versus difference between incoming and outgoing angles $\Delta\theta = \theta_{out} - \theta_{in}$. Green dots: experimental data points. Connected red circles: mean values and standard deviations calculated via binning θ_{in} at every $10°$. Blue straight lines: non-interaction lines to guide the eye. All the 2214 analyzed binary collision events are shown in (**a**), but events with $|\Delta\theta| > 40°$ are not shown in (**b**) for visibility

$\theta_{out} \simeq 0°$ or $180°$. In Fig. 3.11b, we show that the difference between incoming and outgoing angles $\Delta\theta = \theta_{out} - \theta_{in}$ is on average negative for $\theta_{in} < 90°$ and positive for $\theta_{out} > 90°$, which are characteristic of nematic alignment. We note also that our setup allows a significant fraction of events where bacteria cross each other undisturbed (Fig. 3.12a). (On the other hand, we recorded no events where alignment occurs *without* collision, ruling out long-range hydrodynamic effects.) We believe that this weak nematic alignment only due to the weak exclusive volume interactions makes our system even closer to the Vicsek-style models where strong noise allow for non-alignment or even disalignment, which is impossible in strictly two-dimensional experiments [13, 14, 17]. Such 'quasi-two-dimensionality' might play a crucial role in our experiment.

3.4.3 Image Processing

Before moving on to analysis on number fluctuations and nematic order, we calibrated spatial inhomogeneity of epifluorescence illumination in acquired images in the following procedure as shown in Fig. 3.13.

We subtracted time-averaged dark current images from the obtained images of fluorescent *E. coli* (Fig. 3.13a) and divided them by fluorescent images of a homogeneous fluorophore solution (fluorescein) (Fig. 3.13b) to calibrate the spatial inhomogeneity of the excitation light source. The dark current images were also subtracted from the fluorescent images beforehand. In short,

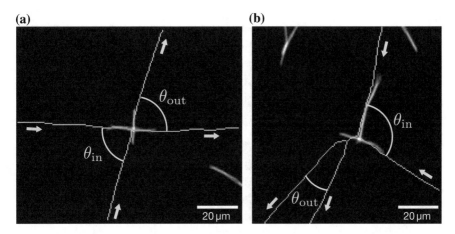

Fig. 3.12 Trajectories of **a** a collision event with almost no interaction and **b** a rare event with an obtuse incoming angle $\theta_{in} > 90°$ and an acute outgoing angle $\theta_{out} < 90°$. Such non-interacting/disaligning events make the system even similar to the Vicsek-style models

$$\text{(calibrated image)} = \frac{\text{(fluorescent } E.\ coli \text{ image)} - \text{(dark current image)}}{\text{(fluorophore image)} - \text{(dark current image)}}. \quad (3.1)$$

Thus we obtained intensity-calibrated images (Fig. 3.13c).

To ensure our analysis methods, we applied the same image processing procedures and the same analysis to both the ordered phase (Fig. 3.8d) and the disordered phase (Fig. 3.14).

3.4.4 Existence of True Long-Range Order

Because we have obtained the ordered phase, we would like to quantify number fluctuations and to know whether the system exhibits giant number fluctuations (GNF) or not. However, as we have stressed many times, the genuine GNF associated with the Nambu-Goldstone modes in broken symmetry states predicted by Toner, Tu, Ramaswamy et al. should be discussed after confirming the presence of long-range order.

Since it is very difficult to determine the polarity θ of each bacterium from the acquired images at such large concentration, a direct estimate of the nematic order parameter $Q = |\langle e^{2i\theta} \rangle|$ previously used even in experiments [32] is out of reach, and we instead introduced the 'structure tensor' method used previously, e.g., for measuring the orientation of collagen fibers in tissues [33]. Specifically, given an intensity-calibrated image $f(x, y)$, one calculates the following tensor over a given region of interest (ROI):

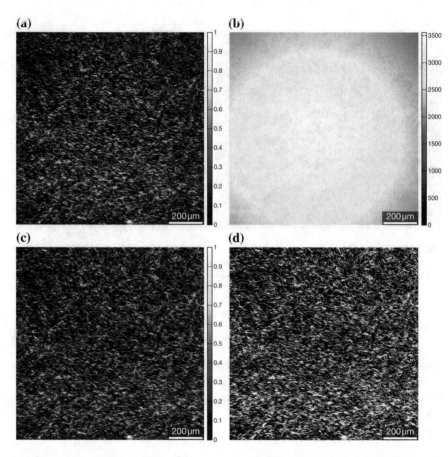

Fig. 3.13 a Raw image after the dark current subtraction. Intensity is adjusted in the range [0, 1] and contrast is enhanced for visibility. This image corresponds to the numerator in Eq. (3.1). **b** Intensity distribution of excitation light source for fluorescent microscopy. Fluorescence intensity of a homogeneous fluorophore solution was obtained by the 12-bit camera and then after dark current subtraction it is used as the denominator of Eq. (3.1). **c** Calibrated image. Intensity is adjusted in the range [0, 1] and contrast is enhanced for visibility. **d** Binarized image used for analysis on number fluctuations

$$J = \begin{bmatrix} \langle \partial_x f, \partial_x f \rangle & \langle \partial_y f, \partial_x f \rangle \\ \langle \partial_x f, \partial_y f \rangle & \langle \partial_y f, \partial_y f \rangle \end{bmatrix} \qquad (3.2)$$

where $\langle g, h \rangle = \iint_{\text{ROI}} g\,h\,\mathrm{d}x\,\mathrm{d}y$. The eigenvalues $\lambda_{\min}$ and $\lambda_{\max}$ of J then give an estimate of the scalar nematic order parameter, called the 'coherency parameter',

$$C \equiv \frac{\lambda_{\max} - \lambda_{\min}}{\lambda_{\max} + \lambda_{\min}}, \qquad (3.3)$$

Fig. 3.14 Whole field of view of the disordered phase at low density. This movie was used for verifying our analysis methods and for comparison with the ordered phase

whereas the eigenvector corresponding to λ_{min} gives the orientation of the global nematic order in the ROI.

We have measured the nematic order parameter $\langle C \rangle$ for square ROIs of various area S, where the average is taken over both space and time. We changed the box size S and moved the boxes in such a way that each box does not overlap each other in order to ensure that all the pixels are used only once for the calculation. In the disordered phases observed either at low density or at high density but in a thicker layer of fluid, we find that $\langle C \rangle \sim 1/\sqrt{S}$, the same behavior as the conventional nematic order parameter $Q = |\langle e^{2i\theta} \rangle|$ in the case of finite spatial correlation length (Fig. 3.15). In the ordered regime observed at the high density and with the thin apparatus, on the other hand, we observe no topological defects and very slow decay of the nematic order parameter (red circles in Fig. 3.15a). The nematic order persists more than millimeters. As shown by the curvature of the *log-log* plot in Fig. 3.15b, this decay is *slower than a power law*. This is the signature of true long-range order, and is distinctively different from short-range exponential correlations found in the coexisting clustering phases of the previous bacterial experiments [13, 14]. As a matter of fact, an excellent fit of the data is an algebraic approach to some finite

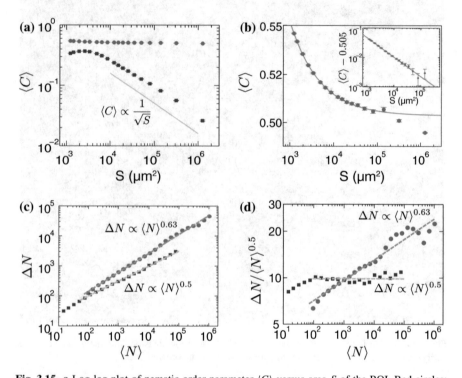

Fig. 3.15 **a** Log-log plot of nematic order parameter $\langle C \rangle$ versus area S of the ROI. Red circles: the globally nematically ordered state at high density in a very thin sample shown in Fig. 3.8c, d. Blue squares: the disordered state at low density shown in Fig. 3.14. Cyan solid line: slope of exponent -0.5 as a guide to the eye. The data of the disordered state deviate from $\langle C \rangle \propto 1/\sqrt{S}$ in small S because the number of ROIs without any bacteria ($C \simeq 0$) increases. The nematic order $\langle C \rangle$ stays at high values over the whole field of view in the ordered state (the red data). **b** Same data as in (**a**) for the ordered state in a magnified range on the log-log scale. The last 3 points were excluded from the fit because they are not reliable due to longer correlation times and inhomogeneities at such large-scales. Inset: same data from which the estimated asymptotic value of $C_\infty = 0.505$ has been subtracted (log-log scale). The curvature in this log-log plot indicates *slower decay than a power law*. Magenta solid lines: fit $\langle C \rangle = C_\infty + kS^\beta$ with $C_\infty = 0.505$, $\beta = -0.66$, and $k = 4.6$. Error bars in (**a**) and (**b**): standard error. **c**, **d** Scaling of number fluctuations ΔN vs $\langle N \rangle$, and $\Delta N/\sqrt{\langle N \rangle}$ vs $\langle N \rangle$ on the log-log scale respectively. Blue squares: normal fluctuations in the disordered, low-density phase. Red circles: anomalous, 'giant' fluctuations recorded in the high-density nematically-ordered state of Fig. 3.8c, d. ΔN of red data points in (**c**) are multiplied by 2 and shifted to avoid overlapping with blue data points. Cyan dashed line: normal fluctuations $\Delta N \propto \langle N \rangle^{0.5}$ as a guide to the eye. Magenta dashed line: fitted curve $\Delta N \propto \langle N \rangle^{0.63}$ for the ordered state. A single filamentous bacterium corresponds to approximately $N \sim 100$ pixels

asymptotic value, $\langle C \rangle - C_\infty \sim S^\beta$, with $C_\infty = 0.505$ and $\beta = -0.66$ (Fig. 3.15b). Similar finite-size scaling was found in the Vicsek-style self-propelled rods model studied in [5].

From the discussion above, we can conclude that nematic order persists even in the large system size limit, which means the existence of true long-range order.[4]

3.4.5 Existence of Giant Number Fluctuations

Because we have already checked that the 'ordered' state of our experimental system is indeed deep in the ordered regime with true long-range order without any clusters, now we can safely discuss GNF genuinely in the sense of the Vicsek universality class.

To quantify number fluctuations, instead of directly detecting each bacterium (again a difficult task), we binarized our images using the commonly-used Otsu's method [34] and counted, in each square ROI centered at the field of view, the number of pixels $N(t)$ covered by bacteria at time t in the same way as in Fig. 2.5a. We calculated a binarization threshold for each frame by applying Otsu's method with MATLAB and used their average value. The boxes used for calculating $N(t)$ were all centered at the field of view in order to avoid incorporating the possible spatial inhomogeneity of the setup into uninteresting 'number fluctuations'. The binarization process has the advantage of correcting for the slight differences in intensity resulting from variations of the height of bacteria or fluctuations of the excitation light intensity (Fig. 3.13d).[5] On the other hand, it leads to small systematic underestimates in the case of overlapping cells. In this analysis, a single filamentous bacterium corresponds to approximately $N \sim 100$ pixels.

We calculated the standard deviations $\Delta N = \sqrt{\langle (N(t) - \langle N \rangle)^2 \rangle}$ (all averages over time) for square ROIs of various sizes. The results are shown in Fig. 3.15c, d. In the disordered phase, we estimate its number fluctuations $\Delta N \sim \langle N \rangle^\alpha$ with the exponent $\alpha = 0.511(12)$ from a linear regression analysis. Here the uncertainty means 95% confidence level estimated from Student's t-test. This is consistent with normal fluctuations $\Delta N \sim \langle N \rangle^{0.5}$ as we expected, which ensures the validity of our experimental analysis. On the other hand, in the dense nematically-ordered phase, we estimate $\Delta N \sim \langle N \rangle^\alpha$ with $\alpha = 0.632(14) > 0.5$ (the uncertainty: 95% confidence level), i.e. anomalous, giant number fluctuations.

---

[4]While, of course, on much larger (inaccessible) scales, this order may break down due to experimental limitations such as imperfectness or boundaries of our setup, our observation done in the area with true long-range order far away from boundaries can extract bulk properties that we are interested in.

[5]When we did not apply binarization and just used calibrated intensity, we could not obtain normal fluctuations with the exponent $\alpha = 0.5$ but *spurious giant* fluctuations with $\alpha > 0.5$ even in the disordered state.

3.4.6   Interpretation of True Long-Range Order and GNF

So far, we have confirmed that our experimental system exhibit giant number fluctuations (GNF) in the long-range ordered nematic phase, which gives the first experimental realization of the Toner-Tu-Ramaswamy phases. However, because we estimated the exponent α in our system by binarization and counting the pixels covered by bacteria, the estimated exponent $\alpha = 0.632(14)$ may not be so precise and cannot simply be compared with theoretical predictions. The important point here is that we obtained the statistically significantly larger exponent α in the ordered phase than in the disordered phase over 3–4 decades.

Our system can be seen as a collection of self-propelled rods without velocity reversals that align nematically. It should thus be compared a priori to the Vicsek-style self-propelled rods model (polar particles with nematic interactions) studied in [5]. Indeed, this model was numerically shown to have true long-range nematic order over all the numerically tested scales, as well as GNF with a scaling exponent $\alpha \simeq 0.75$. Our experimental findings are thus in full qualitative if not quantitative agreement with [5]. Our estimate of α is somewhat smaller than that of this numerical study, but this could be ascribed to excluded volume effects, which, after all, rule most if not all interactions. Excluded volume gives finite upper bound to local density, although local density in the Vicsek-style models can take arbitrary positive values due to the absence of excluded volume of point-like particles.

Furthermore, the theoretical predictions are also not exact. As we have detailed in Chap. 2, the value of α in Toner-Tu-Ramaswamy orientationally-ordered phases is still the matter of debate, even in the case of polar flocks in the original Vicsek model. The value predicted by the Toner-Tu theory in 1995–1998 [7, 8], $\alpha = \frac{4}{5}$ in spatial dimension $d = 2$, is not exact, as originally claimed by Toner's later reanalysis in 2012 [10], and it was only approximately confirmed numerically on the original Vicsek model in [2, 35]. For 'pure' active nematics (apolar particles with fast velocity reversals), the latest numerical estimate of α is again around $\frac{4}{5}$ [6], in contradiction with the linear hydrodynamic theory by Ramaswamy et al. [11]. Here and in [5] a slightly smaller value of α was again found.

3.4.7   Correlation Functions

To further confirm theoretically predicted properties of the Toner-Tu-Ramaswamy phases, we moved on to compare correlation functions in our system with those of theoretical studies. As we have seen in Sect. 2.3.2, the Toner-Tu theory predicts long-range algebraic correlations with characteristic exponents associated with the Nambu-Goldstone mode. Unfortunately, to the best of our knowledge, there are still no analytical results on hydrodynamic equations for the Vicsek-style self-propelled rods, whose symmetry is the same as that of our experimental system. However, we can still try to compare our results with the predictions by the Toner-Tu theory, because both the Vicsek model and the self-propelled rods model have true long-range order in their ordered states and we can expect similar phenomenology (see Fig. 2.11 and Sect. 2.4.2). Again, we still need to be careful that the Toner-Tu theory is not exact, because the derived analytical results are based on the equations without the anisotropic pressure P_2 term that has to be considered as we noted in Sect. 2.3.1 [10].

Correlations of director fluctuations

We measured correlations in the director n of the nematic phase in our experiment. From the structure tensor analysis, we have the local director field n and thus we can calculate the two-point correlation function [36] of director fluctuations $\delta n_\perp = n - n_0$,

$$C_{\delta n_\perp}(R) := \langle \langle \delta n_\perp(r, t) \delta n_\perp(r + R, t) \rangle_r \rangle_t , \qquad (3.4)$$

where n_0 is the global director obtained by spatially averaging n and $\delta n_\perp$ is a signed norm of $\delta n_\perp$, which is shown in Fig. 3.16a, b. Details of calculation of $C_{\delta n_\perp}(R)$ can be found in Appendix Sect. 3.6.2. The counterpart of this correlation function of director fluctuations $\delta n_\perp(r, t)$ in the Toner-Tu theory is the correlation function of velocity fluctuations $C_C(R)$ defined in Eq. (2.22), and the prediction in $d = 2$ was given in Eq. (2.36) as,

$$C_C(R) = \langle v_\perp(r + R, t) \cdot v_\perp(r, t) \rangle \propto \begin{cases} R_\perp^{2\chi} = R_\perp^{-2/5} = R_\perp^{-0.4} \\ R_\parallel^{2\chi/\zeta} = R_\parallel^{-2/3} = R_\parallel^{-0.667}, \end{cases} \qquad (3.5)$$

As we have already remarked on analysis with the finite field of view in Sect. 2.3.3, the constraint $\langle \delta n_\perp(t, r) \rangle_r = 0$ makes the experimentally estimated correlation function $C_{\delta n_\perp}(R)$ become negative at certain distances. This makes extracting asymptotic algebraic behavior from experimental data quite difficult. Nonetheless, we have strived to extract its asymptotic behavior and the exponents. In both longitudinal and transverse directions against n_0, $C_{\delta n_\perp}(R)$ decays algebraically from the cell length of $\sim 20\ \mu m$ up to the scale where the inhomogeneities of the setup and such effects of the finite field of view are more pronounced (Fig. 3.16b). Linear fitting on the log-log scale gives us the estimates of the exponents,

$$C_{\delta n_\perp}(R) = \langle \langle \delta n_\perp(r, t) \delta n_\perp(r + R, t) \rangle_r \rangle_t \propto \begin{cases} R_\perp^{-0.405(14)} \\ R_\parallel^{-0.333(6)}, \end{cases} \qquad (3.6)$$

where fitting ranges are [60 μm, 131 μm] for the transverse direction and [60 μm, 250 μm] for the longitudinal direction respectively, and the uncertainty means 95% confidence level estimated from Student's t-test.

By equating Eqs. (3.5) and (3.6), we obtain experimental estimates of the exponents χ_{exp} and ζ_{exp} as,

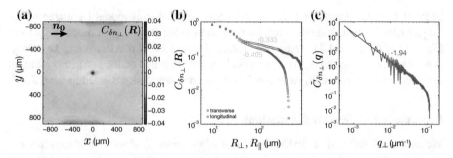

Fig. 3.16 **a** Colormap of the correlation function $C_{\delta n_\perp}(\boldsymbol{R})$. The director $\boldsymbol{n_0}$ is aligned to the x-direction. **b** Log-log plot of the correlation function $C_{\delta n_\perp}(\boldsymbol{R})$. Blue circles: the transverse direction (perpendicular to $\boldsymbol{n_0}$). Red squares: the longitudinal direction (along $\boldsymbol{n_0}$). Linear fitting gives the exponent $-0.405(14)$ for transverse direction (cyan solid line) and $-0.333(6)$ for the longitudinal direction (yellow solid line) respectively. **c** The Fourier transformed correlation function of director fluctuations $\tilde{C}_{\delta n_\perp}(\boldsymbol{q})$ in the transverse direction. It exhibits algebraic behavior over a wide range. Red line: the result of linear fitting gives the exponent $-1.94(11)$. Wavenumber $q_\perp$ is not multiplied by 2π

$$2\chi_{\text{exp}} \sim -0.405(14) \Rightarrow \chi_{\text{exp}} \sim -0.202(7) \quad (\text{prediction: } \chi = -1/5 = -0.2),$$

(3.7)

$$\frac{2\chi_{\text{exp}}}{\zeta_{\text{exp}}} \sim -0.333(6) \Rightarrow \zeta_{\text{exp}} \sim 1.22(5) \quad (\text{prediction: } \zeta = 3/5 = 0.667). \quad (3.8)$$

The experimentally obtained exponent χ_{exp} is nicely consistent with the predicted value in the Toner-Tu theory [7, 8]. However, the anisotropy exponent ζ_{exp} deviates from the predicted value. Therefore, the correlations of director/velocity fluctuations in our experimental system and in the Toner-Tu theory match well only in the transverse direction.

We also investigated this correlation function in the Fourier space,

$$\tilde{C}_{\delta n_\perp}(\boldsymbol{q}) = \int d^2\boldsymbol{r}\, C_{\delta n_\perp}(\boldsymbol{R}) e^{i\boldsymbol{q}\cdot\boldsymbol{R}}. \quad (3.9)$$

From the experimental data in the transverse direction, we obtained an algebraic correlation as shown in Fig. 3.16c. We applied linear fitting in $q_\perp < 0.03\ \mu\text{m}^{-1}$ and obtained the scaling,

$$\tilde{C}_{\delta n_\perp}(\boldsymbol{q}) \sim q_\perp^{-1.94(11)}, \quad (3.10)$$

where the uncertainty means 95% confidence level estimated from Student's t-test.

The behavior of $\tilde{C}_{\delta n_\perp}(\boldsymbol{q})$ should be compared with the correlation function in the Toner-Tu theory defined in Eq. (2.29) and predicted for $d = 2$ in Eq. (2.37) as,

$$C_{ij}(\boldsymbol{q}) = \langle v_i^\perp(\boldsymbol{q}, t) v_j^\perp(-\boldsymbol{q}, t) \rangle \sim q_\perp^{1-d-\zeta-2\chi} = q_\perp^{-6/5} = q_\perp^{-1.2}. \quad (3.11)$$

The experimentally obtained exponent, $-1.94(11)$, is significantly different from the Toner-Tu prediction, -1.2. However, if we substitute the experimentally obtained values χ_{exp} and ζ_{exp} into the exponent in Eq. (3.11), we obtain

$$1 - d - \zeta_{\text{exp}} - 2\chi_{\text{exp}} = -1.81(5), \tag{3.12}$$

which overlaps with the experimentally obtained exponent $-1.94(11)$ of $\tilde{C}_{\delta n_\perp}(\boldsymbol{q})$ within the range of uncertainty. This relation, derived under some assumptions by Toner and Tu, is a kind of hyperscaling relation among scaling exponents. Our experimental data suggest the possible validity of this hyperscaling relation, although there is no rigorous proof in the Tone-Tu theory and our system's symmetry is different from that of the Toner-Tu theory.

We also note that both the *linear* approximation of the Toner-Tu equations [8, 9] and the *linear* theory on hydrodynamic equations of active nematics (apolar particles with nematic interactions) by Ramaswamy et al. [9, 11] yield the following scalings on $C_{ij}(\boldsymbol{q})$ and $\tilde{C}_{\delta n_\perp}(\boldsymbol{q})$ respectively,

$$C_{ij}(\boldsymbol{q}) \sim q^{-2}, \quad \tilde{C}_{\delta n_\perp}(\boldsymbol{q}) \sim q^{-2}. \tag{3.13}$$

The same $\sim q^{-2}$ behavior can also be observed in *equilibrium* nematic liquid crystals [36].[6] In this sense, our global nematic phase with fluctuations $\tilde{C}_{\delta n_\perp}(\boldsymbol{q}) \sim q_\perp^{-1.94(11)}$ might be said to be 'ideally nematic'. Nonetheless, we need to be careful about whether the exponent of our system is different from -2 or not, because this exponent should be connected to other exponents such as α for number fluctuations. However, unfortunately, there is no analytical calculation on correlation functions in active nematic phases except for the *linear* theory for active nematics [11], so we have to develop such theories in the future.

Density correlation functions

We calculated equal-time density correlation functions, or the structure factors $S(\boldsymbol{q})$. The corresponding definitions and predictions in the Toner-Tu theory were given in Eqs. (2.28) and (2.38) as,

$$C_\rho(\boldsymbol{q}) = S(\boldsymbol{q}) = \langle |\rho(\boldsymbol{q}, t)|^2 \rangle \sim q_\perp^{1-d-\zeta-2\chi} = q_\perp^{-6/5} = q_\perp^{-1.2}. \tag{3.14}$$

Because we cannot know the exact positions and the exact density field ρ of bacteria from the images in such high concentration, we estimated the local density from the binarized images, which are the same images used for the GNF analysis above. From each binarized black/white (0 or 1) image $\text{BW}(\boldsymbol{r}, t)$ at time t, we obtained an estimate of the equal-time density correlation functions in the Fourier space as,

---

[6]In equilibrium liquid crystals and in active nematics, this $\sim q^{-2}$ fluctuations of director arising from the Nambu-Goldstone mode break the global order, leading to only quasi-long-range order.

$$\tilde{C}_\rho(\boldsymbol{q}) \propto \tilde{C}_{\mathrm{BW}}(\boldsymbol{q}) = \left\langle \left| \int \mathrm{BW}(\boldsymbol{r}, t) e^{-i\boldsymbol{q}\cdot\boldsymbol{r}} \mathrm{d}\boldsymbol{r} \right|^2 \right\rangle_t , \qquad (3.15)$$

where the constant of proportionality is required due to the binarization process. In the actual computational process, in order to reduce artifacts caused by non-periodicity of $\mathrm{BW}(\boldsymbol{r}, t)$ in the finite field of view, we first removed the DC ($\boldsymbol{q} = \boldsymbol{0}$) components by subtracting the mean value of the images and then multiplied a window function to $\mathrm{BW}(\boldsymbol{r}, t)$ before applying Fast Fourier Transform. As a window function, we chose the two-dimensional hanning window $\mathrm{Hann}(x, y)$ defined over $0 \le x \le 1$ and $0 \le y \le 1$ as,

$$\mathrm{Hann}(x, y) = 0.5(1 - \cos 2\pi x) \times 0.5(1 - \cos 2\pi y). \qquad (3.16)$$

After obtaining the Fourier transform, we corrected their amplitude because it is modified by multiplication with the window function.

As we have explained in Sect. 3.4.5, calibrated fluorescent images $f(\boldsymbol{r}, t)$ of bacteria still contains unwanted intensity fluctuations. Therefore, we used binarized images as the estimate of the bacterial density field. Of course, this binarization process dismisses overlapping bacteria and underestimates the local density, so this approximation might lead to wrong results on the density correlation functions at length scales comparable to the size of a single bacterium. However, this approximation cannot affect the density correlation at the large scale where we are interested in. We have checked with our experimental data that the binarization process affects only at short length scales (large wavenumbers) not at large length scales (small wavenumbers).

The calculated $\tilde{C}_{\mathrm{BW}}(\boldsymbol{q})$ for both the ordered state and the disordered state is shown in Fig. 3.17. In the disordered state, $\tilde{C}_{\mathrm{BW}}(\boldsymbol{q})$ both the x- and the y-directions in the images exhibit the same behavior due to the isotropy of the system. It converges to a finite value in $\boldsymbol{q} \to \boldsymbol{0}$ limit. This convergence is characteristic of normal fluctuations with $\alpha = 0.5$. This result is consistent with our results presented in Sect. 3.4.5. On the other hand, in the ordered state, algebraic behavior and divergence at small $\boldsymbol{q}$ were found both in the transverse and in the longitudinal direction with respect to the global order, although the spatial inhomogeneity in the experimental setup restrict the ranges in which we can see the algebraic behavior. The divergence in the $\boldsymbol{q} \to \boldsymbol{0}$ limit implies the existence of the genuine GNF, as we have seen in Eq. (2.34). Linear fitting was applied to extract the exponents. We obtained,

$$\tilde{C}_{\mathrm{BW}}(\boldsymbol{q}) \sim \begin{cases} q_\perp^{-1.2011(72)} & \text{for the transverse direction} \\ q_\parallel^{-1.780(13)} & \text{for the longitudinal direction} \end{cases} \qquad (3.17)$$

Here, fitting ranges are $[5 \times 10^{-2}\ \mu\mathrm{m}^{-1}, 3 \times 10^{-1}\ \mu\mathrm{m}^{-1}]$ for the transverse direction and $[2 \times 10^{-2}\ \mu\mathrm{m}^{-1}, 5 \times 10^{-1}\ \mu\mathrm{m}^{-1}]$ for the longitudinal direction respectively, and the uncertainty means 95% confidence level estimated from Student's t-test.

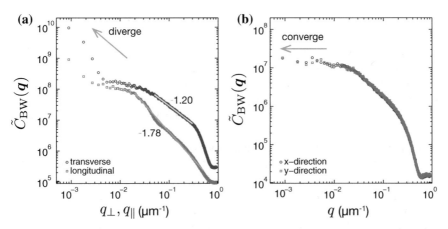

Fig. 3.17 The Fourier-transformed equal-time density correlation function $\tilde{C}_{BW}(q)$. This corresponds to $C_{\rho}(q)$ in the Toner-Tu theory, or the structure factor $S(q)$. **a** $\tilde{C}_{BW}(q)$ in the ordered phase. Blue circles: transverse to the global director n_0. Red squares: along the global director n_0. Algebraic behavior is observed in both directions, which indicates giant number fluctuations with $\alpha > 0.5$. The estimated exponents are $-1.2011(72)$ and $-1.780(13)$ for the transverse and the longitudinal directions respectively. Cyan and green solid lines are results of linear fittings. **b** $\tilde{C}_{BW}(q)$ in the disordered phase. Blue circles: the x-direction of the images. Red squares: the y-direction of the images. $\tilde{C}_{BW}(q)$ converges to a finite value in the limit of $q \to 0$, which means normal number fluctuations with $\alpha = 0.5$. Wavenumbers, $q_{\perp}, q_{\parallel}, q$, are not multiplied by 2π

The Toner-Tu prediction in Eq. (3.14) and our results in Eq. (3.17) coincide surprisingly well in the transverse direction. If we believe the Toner-Tu theory, this exponent -1.2 in the transverse direction of the Fourier-transformed density correlation function $C_{\rho}(q)$ corresponds to the GNF exponent $\alpha = 0.8$, as we can see from Eqs. (2.34) and (2.35). In this sense, we might be able to argue that our ordered phase can actually have the GNF exponent $\alpha \simeq 0.8$ instead of our experimental observation $\alpha = 0.63$, if we could perform ideal experiments by removing all the experimental difficulties such as the inhomogeneity of the setup. Furthermore, this estimate of α from the correlation function $\tilde{C}_{BW}(q)$ can be more robust against the binarization process than the direct estimate obtained by counting the pixels covered by bacteria, the method we used in Sect. 3.4.5. However, we need to be careful about interpretation of this results due to the limited range of algebraic behavior both in the GNF (Fig. 3.15c,d) and in the correlation function (Fig. 3.17a).

We note that the exponent in the longitudinal direction is difficult to compare with the Toner-Tu theory due to the complicated functional form given by Toner and Tu [8, 9].

Short summary of correlation functions

The algebraic correlations we have seen above are manifestations of the Nambu-Goldstone mode in the ordered states with broken rotational symmetry, which are closely connected to the emergence of GNF [7–11]. In particular, these algebraic *scale-free* correlations support our claim that the ordered state obtained in our experiments are *homogeneous without any clusters*, excluding the possibility of the existence of clusters with characteristic length scales. Therefore, the ordered state of our system is indeed an example of the Toner-Tu-Ramaswamy phase.

Although the predictions in the Toner-Tu theory relies on some assumptions and also may not be applicable to our system due to the different symmetries, some of our experimental data are still consistent with the Toner-Tu predictions. The asymptotic exponents of the correlation functions of director fluctuations in real space $C_{\delta n_\perp}(\boldsymbol{R})$ turned out to be the same only in the transverse direction as the prediction on the correlation function of velocity fluctuations $C_C(\boldsymbol{R})$ in the Toner-Tu theory on the *polar* flocks, but they do not coincide in the longitudinal direction. This means that our experimental system and the Toner-Tu theory share the exponent χ but not the anisotropy exponent ξ. The correlation function of director fluctuations in the Fourier space $\tilde{C}_{\delta n_\perp}(\boldsymbol{q})$ has the exponent $-1.94(11)$, which is close to the equilibrium nematics or the prediction by Ramaswamy's linear theory on active nematics rather than that in the Toner-Tu theory. However, we have still confirmed that the hyperscaling relation predicted in the Toner-Tu theory might hold within the experimental uncertainty. The Fourier transformed density correlation function $\tilde{C}_{\mathrm{BW}}(\boldsymbol{q})$, or the structure factor $S(\boldsymbol{q})$, exhibits the same exponent as in the Toner-Tu theory in the transverse direction.

We have examined the correspondence between our experimental results and the existing theories, but we still need to seek for more precise analytical results for collective motion, especially for self-propelled rods (polar particles with nematic interactions), to better understand our experimental systems and to find out the validity of hydrodynamic theories. Our experimental results challenge the existing theories by raising many questions, and give clues for developing better theories.

3.4.8   Boltzmann Approach

Idea and Claim

So far, we have mainly examined the statistical properties of the ordered state. Here we focus on the transition from the disordered to the ordered phases from the microscopic point of view. Because we have collected the statistics on binary collision events, our next question is whether we can understand the transition to the nematically-ordered state from the microscopic collision dynamics or not. To clarify this point, we employed a kinetic approach using the Boltzmann equation, and investigated the linear stability of the isotropic disordered state.

As a result of the analysis we present below, it has turned out that the disordered, isotropic, homogeneous phase is linearly stable on the kinetic level, hence binary collisions alone cannot account for the transition to neither a polar phase nor a nematic phase. The weak alignment interactions induced by collisions are not sufficient for transition to the ordered phase in terms of the Boltzmann equation, which apparently contradicts with our experimental observation of the ordered phase. This apparent conundrum is resolved by the fact that, in our high-density experiment, isolated binary collisions almost never happen. As we can see from our experimental images (Fig. 3.8), the number density of bacteria is so high that every bacterium is almost always in contact with at least another bacterium. Therefore, assumptions of dilute limit and molecular chaos—correlated velocities of two colliding bacteria—in the Boltzmann approach are too strong to explain the experiments, which suggests the importance of multi-particle collisions in the transition to the ordered phase. A similar argument can be found in a motility assay experiment [37].

Formalism

We start from the Boltzmann equation for self-propelled particles [37–40].[7] It can describe the time evolution of the one-particle distribution function $f(r, \theta, t)$ under the assumption of dilute limit, where r is the particle position, θ is the particle orientation, and t is time. It can be written as,

$$\frac{\partial}{\partial t} f(r, \theta, t) + v_0 \hat{v}(\theta) \cdot \nabla f(r, \theta, t) = D_\theta \frac{\partial^2}{\partial \theta^2} f(r, \theta, t) + C[f^{(2)}], \qquad (3.18)$$

where v_0 is the speed of the particles which is assumed to be a constant here, $\hat{v}(\theta)$ is a unit vector directing to the θ-direction, D_θ is the angular diffusion constant, and $C[f^{(2)}]$ is the collision integral. The collision integral can be calculated from the two-particle distribution $f^{(2)}(r, \theta_1, \theta_2, t)$, which can be rewritten using the one-particle distribution as $f^{(2)}(r, \theta_1, \theta_2, t) = f(r, \theta_1, t) f(r, \theta_2, t)$ in the absence of correlations (molecular chaos assumption),[8] where θ_j is the orientation of the j-th particle. The collision integral can be written as,

$$C[f^{(2)}(r, \theta_1, \theta_2, t)] = -\int_{-\pi}^{\pi} d\theta' \Gamma(\theta_{12}) f^{(2)}(\theta, \theta')$$

$$+ \int_{-\pi}^{\pi} d\theta_1 \int_{-\pi}^{\pi} d\theta_2 \frac{1}{2} \sum_{j=1}^{2} \Gamma(\theta_{12}) f^{(2)}(\theta_1, \theta_2) \int_{-\infty}^{+\infty} d\eta_j \, p_j(\eta_j|\theta_{12}) \overline{\delta}(\theta_j + \eta_j - \theta),$$

$$(3.19)$$

where θ_{12} is the signed incoming angle or the angle difference of the two colliding particles, η_j is the difference between the orientations of the j-th particle before and after the collision, and $\overline{\delta}$ is a generalized Dirac delta imposing that the argument is equal to zero modulo 2π: $\overline{\delta}(\theta) = \sum_{m=-\infty}^{+\infty} \delta(\theta + 2\pi m)$. Here, $\Gamma(\theta_{12})$ is a collision kernel describing the collision rate, which depends solely on the incoming angle θ_{12}

---

[7]In the following, we neglect the positional diffusion term, which is small in our experiment and does not affect the linear stability analysis.

[8]Of course, at the onset of collective motion, there should be a strong correlation in the angles θ_1 and θ_2 of the two colliding particles. Taking this into account, in [37, 38], the angular correlations was given in the form of $f^{(2)}(r, \theta_1, \theta_2, t) = (1 + A/|\theta_{12}|) f(r, \theta_1, t) f(r, \theta_2, t)$. However, this dependence cannot be justified by experiments and, furthermore, this does not modify the obtained results qualitatively.

due to a global rotational invariance of the system. If we assume that our filamentous bacteria can be treated as rods with the diameter d, the length L, and the aspect ratio $\xi = L/d$, we can theoretically derive a functional form of $\Gamma(\theta_{12})$, just by a geometrical construction called 'the Boltzmann scattering cylinder' [41], as,

$$\Gamma(\theta_{12}) = 4dv_0 \left| \sin\left(\frac{\theta_{12}}{2}\right) \right| \left(1 + \frac{\xi - 1}{2}|\sin\theta_{12}|\right). \tag{3.20}$$

For a given incoming angle θ_{12}, the orientation of the j-th particle changes by $\eta_j(\theta_{12})$ with probability $p_j(\eta_j|\theta_{12})\mathrm{d}\eta_j$. We note that our experimental binary collision statistics will be used for calculating $p_j(\eta_j|\theta_{12})\mathrm{d}\eta_j$ later.

To analyze the linear stability of the disordered phase on the kinetic level, we apply an angular Fourier expansion defined as

$$f(\boldsymbol{r}, \theta, t) = \frac{1}{2\pi} \sum_{k=-\infty}^{+\infty} \hat{f}_k(\boldsymbol{r}, t)\mathrm{e}^{-ik\theta}, \tag{3.21}$$

where the Fourier coefficients $\hat{f}_k(\boldsymbol{r}, t)$ are defined as,

$$\hat{f}_k(\boldsymbol{r}, t) = \int_{-\pi}^{\pi} \mathrm{d}\theta \, f(\boldsymbol{r}, \theta, t)\mathrm{e}^{ik\theta}. \tag{3.22}$$

Note that $\hat{f}_{-k} = \hat{f}_k^*$ for all k, where $*$ denotes the complex conjugate. It is important to understand the physical meaning of each Fourier coefficients $\hat{f}_k(\boldsymbol{r}, t)$, especially for $k = 0, 1, 2$. Because $\hat{f}_0(\boldsymbol{r}, t)$ is obtained by integrating out the angular dependence of the one-particle distribution function $f(\boldsymbol{r}, \theta, t)$, it is equal to the density field $\rho(\boldsymbol{r}, t)$. The higher Fourier modes $\hat{f}_k(\boldsymbol{r}, t)$ ($k \geq 1$) can be regarded as 'order parameter fields' for corresponding spontaneous symmetry breaking from isotropic states to k-fold rotational symmetry states. In particular, $\hat{f}_1(\boldsymbol{r}, t)$ and $\hat{f}_2(\boldsymbol{r}, t)$ encode the momentum field and the tensorial nematic order parameter field respectively as,

$$\rho\boldsymbol{P} = \begin{pmatrix} \mathrm{Re}\,\hat{f}_1 \\ \mathrm{Im}\,\hat{f}_1 \end{pmatrix}, \quad \rho\boldsymbol{Q} = \frac{1}{2}\begin{pmatrix} \mathrm{Re}\,\hat{f}_2 & \mathrm{Im}\,\hat{f}_2 \\ \mathrm{Im}\,\hat{f}_2 & -\mathrm{Re}\,\hat{f}_2 \end{pmatrix}. \tag{3.23}$$

Here, $\boldsymbol{P}$ is a coarse-grained polarity field with components $P_i = \langle a_i \rangle$ and $\boldsymbol{Q}$ is the coarse-grained traceless tensorial nematic order parameter field with components $Q_{ij} = \langle a_i a_j \rangle - \delta_{ij}/2$, where $\boldsymbol{a}$ is a unit vector representing orientation of each particle and $\langle\,\rangle$ denotes a local average over the distribution $f(\boldsymbol{r}, \theta, t)$. More details of the order parameters are given in Sect. 3.6.3 as Appendix.

As a result of the Fourier expansion, for the k-th Fourier mode $\hat{f}_k$, we obtain,

$$\frac{\partial}{\partial t}\hat{f}_k + \frac{v_0}{2}\left[\frac{\partial}{\partial x}(\hat{f}_{k+1} + \hat{f}_{k-1}) - i\frac{\partial}{\partial y}(\hat{f}_{k+1} - \hat{f}_{k-1})\right]$$

$$= -k^2 D_\theta \hat{f}_k - \frac{4Lv_0}{\pi}\sum_{n=-\infty}^{+\infty}\hat{f}_n \hat{f}_{k-n}[2\mathcal{I}_n - \mathcal{J}_{n,k}], \tag{3.24}$$

where

$$\mathcal{I}_n = \int_{-\pi}^{\pi} d\theta_{12}\ \cos(n\theta_{12})\frac{1}{\xi}\left|\sin\left(\frac{\theta_{12}}{2}\right)\right|\left(1 + \frac{\xi-1}{2}|\sin\theta_{12}|\right), \tag{3.25}$$

$$\mathcal{J}_{n,k} = \mathcal{J}_{n,k}^{(1)} + \mathcal{J}_{n,k}^{(2)}, \tag{3.26}$$

$$\mathcal{J}_{n,k}^{(2)} = \int_{-\pi}^{\pi} d\theta_{12}e^{-in\theta_{12}}G_2(k|\theta_{12})\frac{1}{\xi}\left|\sin\left(\frac{\theta_{12}}{2}\right)\right|\left(1 + \frac{\xi-1}{2}|\sin\theta_{12}|\right), \tag{3.27}$$

$$G_j(k|\theta_{12}) = \int_{-\infty}^{\infty} d\eta_j\ e^{ik\eta_j}p_j(\eta_j|\theta_{12}). \tag{3.28}$$

$G_j(k|\theta_{12})$ is the characteristic function of the probability distribution $p_j(\eta_j|\theta_{12})$. Because $G_1(k|\theta_{12}) = G_2^*(k|\theta_{12})$, the relation $\mathcal{J}_{n,k}^{(1)} = \left(\mathcal{J}_{n,k}^{(2)}\right)^*$ holds, where the asterisk * represents complex conjugation. In the above Eqs. (3.25)–(3.28), the representation of $\Gamma(\theta_{12})$ given in Eq. (3.20) is already substituted.

Because we are interested in whether the transition from the isotropic disordered state to the ordered state can be triggered from the viewpoint of the Boltzmann equation, we investigate the linear stability of the isotropic disordered state. Therefore, we linearize Eq. (3.24) around the solution for the isotropic state, $\hat{f}_0 = \rho_0$ and $\hat{f}_k = 0$ ($|k| \geq 1$), with ρ_0 the global mean density. We obtained the linearized equations for $k = 1, 2$ as,

$$\frac{\partial}{\partial t}\hat{f}_1 = \left[-D_\theta + \frac{Lv_0\rho_0}{\pi}\{\mathcal{J}_{0,1} + \mathcal{J}_{1,1} - 2\mathcal{I}_1 - 2\mathcal{I}_0\}\right]\hat{f}_1, \tag{3.29}$$

$$\frac{\partial}{\partial t}\hat{f}_2 = \left[-4D_\theta + \frac{Lv_0\rho_0}{\pi}\{\mathcal{J}_{0,2} + \mathcal{J}_{2,2} - 2\mathcal{I}_2 - 2\mathcal{I}_0\}\right]\hat{f}_2. \tag{3.30}$$

Whether polar/nematic order emerges or not can be understood by looking at coefficients of $\hat{f}_k$ in the right-hand side of the linearized equations for $k = 1, 2$ respectively. If the coefficients are negative, the Fourier modes $\hat{f}_k = 0$ are linearly stable. If they are positive, the Fourier modes $\hat{f}_k$ grow and thus the isotropic disordered phase is linearly unstable, which results in the polar/nematic states.

Inferring from experimental data

In order to estimate the coefficients in Eqs. (3.29) and (3.30), we need to substitute the experimental values of D_θ, L, v_0, ρ_0, $\mathcal{J}_{0,1}$, $\mathcal{J}_{1,1}$, $\mathcal{J}_{0,2}$, $\mathcal{J}_{2,2}$, $\mathcal{I}_0$, $\mathcal{I}_1$, and $\mathcal{I}_2$. Because D_θ, L, v_0, and ρ_0 are all positive constants and do not encode the interactions through collisions, the most important part in the coefficients is the terms of $\mathcal{J}$ and $\mathcal{I}$ in the brackets. In fact, the experimentally obtained collision statistics enter the $\mathcal{J}$ terms.

Fig. 3.18 Definitions of the signs of the singed incoming angles θ_{12} and the angular difference before and after collisions θ_j

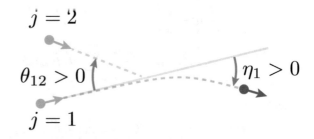

From Eq. (3.25), we have analytical representations of $\mathcal{I}_0$, $\mathcal{I}_1$, and $\mathcal{I}_2$, as,

$$\mathcal{I}_0 = 4 \tag{3.31}$$

$$\mathcal{I}_1 = -\frac{4(-2 + 5\xi)}{9\xi} \tag{3.32}$$

$$\mathcal{I}_0 = \frac{4(-32 + 17\xi)}{225\xi}. \tag{3.33}$$

Next, we calculate the $\mathcal{J}$ terms. From the collision statistics we have obtained in Fig. 3.11, we can construct the probability distribution $p_j(\eta_j|\theta_{12})\mathrm{d}\eta_j$ that appear inside the integral of the definitions of $\mathcal{J}$ in Eqs. (3.26)–(3.28). Although we did not consider the signs of the incoming angles θ_{in} before in Fig. 3.11, here we need that information. So we defined the signs of the signed incoming angles θ_{12} and the angular changes due to collisions η_j as shown in Fig. 3.18. The sign of θ_{12} is positive (negative) when the particle of interest (i.e. $j = 1$) collides with another particle coming from its left (right) side. The sign of η_j is positive in the clockwise direction. The reason for these definitions is just to simplify the computations. Of course, different definitions of the signs do not change the results presented here.

We can extract two sets of the incoming angles θ_{12} and the angle change η_j for the particles $j = 1, 2$, which can be used to increase the statistics. For this purpose, we utilized (i) particle exchange symmetry $p_1(\eta_1|\theta_{12}) = p_2(\eta_2|\theta_{12})$, and (ii) mirror symmetry $p_j(\eta_j| - \theta_{12}) = p_j(-\eta_j|\theta_{12})$. These two symmetries lead to the relation $p_1(\eta_1|\theta_{12}) = p_2(-\eta_2|\theta_{12})$. By using this relation, we can increase the statistics for calculating $p_j(\eta_j|\theta_{12})\mathrm{d}\eta_j$ from the collision events.

By binning the obtained collision statistics at every $5°$ in both θ_{12} and θ_j, we obtained the probability distribution $p_j(\eta_j|\theta_{12})\mathrm{d}\eta_j$ as shown in Fig. 3.19. This distribution is normalized so that $\int p_j(\eta_j|\theta_{12})\mathrm{d}\eta_j = 1$. Then, we apply the Fourier transform to $p_j(\eta_j|\theta_{12})$ to obtain its characteristic function $G_j(k|\theta_{12})$ according to Eq. (3.28). Therefore, by using (3.26)–(3.28), we obtained the values of $\mathcal{J}$ terms as,

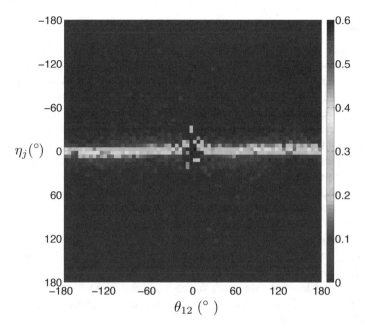

Fig. 3.19 Color plot of the probability distribution $p_j(\eta_j|\theta_{12})\mathrm{d}\eta_j$. It is normalized so that the summation over η_j equals 1 for each θ_{12}. Bin size: 5° for both θ_{12} and η_j

$$\mathcal{J}_{0,1} = 0.0428, \tag{3.34}$$

$$\mathcal{J}_{0,2} = -0.0086, \tag{3.35}$$

$$\mathcal{J}_{1,1} = 0.0251, \tag{3.36}$$

$$\mathcal{J}_{1,1} = 0.0165, \tag{3.37}$$

where we substituted the aspect ratio of bacteria $\xi \simeq 20$, because the length of our bacteria $L \simeq 19\,\mu\mathrm{m}$ and their diameter is smaller than $1\,\mu\mathrm{m}$. The overall results were robust against varying the value of ξ.

Finally, we obtained,

$$\mathcal{J}_{0,1} + \mathcal{J}_{1,1} - 2\mathcal{I}_1 - 2\mathcal{I}_0 = -16.8, \tag{3.38}$$

$$\mathcal{J}_{0,2} + \mathcal{J}_{2,2} - 2\mathcal{I}_2 - 2\mathcal{I}_0 = -8.5, \tag{3.39}$$

both of which are negative. The negativity of these values was robust against variation of the value ξ and the binning size. Therefore, taking into account that the diffusion constant D_θ and $\frac{Lv_0\rho_0}{\pi}$ are both positive, the coefficients of both $\hat{f}_1$ and $\hat{f}_2$ in the linearized Boltzmann equations Eqs. (3.29) and (3.30) are negative, irrespective the values of D_θ, L, v_0, and ρ_0. Hence, both $\hat{f}_1 = 0$ and $\hat{f}_2 = 0$ are linearly stable, which means that neither polar nor nematic order emerges spontaneously from the kinetic arguments presented above.

Short summary of the Boltzmann approach

In conclusion, we cannot account for the transition to the collective phase from the binary collisions inducing weak nematic alignment, which indicates the importance of multi-particle collisions in the collective dynamics of our filamentous bacteria as we have expected. This fact suggests that, in the high-density regime, systematic strong alignment is not required for the emergence of order. Multiple collisions with weak average alignment, repeated in space and time, can suffice. In fact, even if alignment effects are too weak at the binary collision level to produce order (in the 'Boltzmann-like regime'), they can nevertheless lead to order when many-body collisions are taken into account. As a matter of fact, in our high-density case, a given bacterium is almost always in contact with at least one other cell. A rough estimate of the inter-bacterial distance l_{inter}, calculated from the area fraction of our bacteria ~ 0.25 in the ordered state and the number of pixels covered by a single filamentous cells $N \sim 100$, is approximately $l_{inter} \sim 11$ μm, which is indeed shorter than the lengths of our filamentous bacteria. This supports the importance of continuous/multi-particle collisions.

To further understand this point, future work should focus on developing both theoretical tools to describe multi-particle collisions and experimental methods to capture their dynamics.

3.5   Discussion and Conclusion

We have realized a homogeneous nematically-ordered phase of filamentous bacteria swimming in a very thin fluid layer between two solid walls. This phase has turned out to have true long-range order, and in this true long-range ordered phase we confirmed the existence of giant number fluctuations (GNF) and algebraic correlations both in the director fluctuations and in the density field. All of these results signify that the ordered phase in our system is indeed the first experimental realization of the Toner-Tu-Ramaswamy phases and that our system falls into the Vicsek universality class.

To the best of our knowledge, other works reporting the Toner-Tu-Ramaswamy features like GNF either clearly deal with regimes actually disordered [14, 42] and/or are strongly limited by boundaries [43–45]. Here, our system is ordered at least as long as we can observe it, the boundaries are far away, and the interactions, which are essentially due to collisions, are very short-range. Thus our system can undoubtedly be regarded as a Toner-Tu-Ramaswamy phase.

Any existing experiments and observations of bacterial systems have never found long-range order, although theoretical and numerical works on the Vicsek-style models and the hydrodynamic theories predict universality. Even in two-dimensional (2D) numerical simulations on realistic particles with excluded volume, most of them end up clustering or swarming phases without long-range order. Until our experimental realization, there has been no clue for understanding why such 'universality' is not universally found in real flocking systems. Our experiments give insight on necessary conditions for developing the Toner-Tu-Ramaswamy phases.

We speculate that the crucial difference from the other experiments, especially those of bacterial experiments, is *quasi*-two-dimensionality (quasi-2D) of our setup, unlike strictly 2D experiments on agar plates. The quasi-two-dimensionality, and thus adequately weak excluded volume interactions, allow bacteria to cross/overlap during collisions, which is prohibited in strictly 2D systems. In strictly 2D systems, collisions induce formations of clusters because particles cannot escape from the dense regions by crossing other particles and the orientations of clusters fluctuate due to continuous collisions leading to swarming phases [13, 14, 17]. On the other hand, in our quasi-2D setup, bacteria can escape from the dense regions thanks to the other tiny but finite dimension, which prevents cluster formation. As a matter of fact, we observe large fluctuations in the collision statistics (Fig. 3.11) and rare non-interacting or even disaligning events (Fig. 3.12), which make our system even closer to the Vicsek-style interactions than those in other bacterial experiments. Furthermore, such strong confinement kills long-range hydrodynamic interactions that lead to turbulent states in three-dimensional (3D) systems [15–18]. In line with this speculation, future work should focus on experiments with variable dimensionality (2D, quasi-2D, and 3D) of the system and numerical studies on quasi-2D systems with weak volume exclusions. As a matter of fact, very recently, the importance of weak excluded volumes has got supported by a numerical study [46] and an experimental study [47] with variable excluded volumes, and a new nonequilibrium force that stabilizes orientational order in such a confined geometry has also been theoretically proposed [48]. As such, our experiments not only give the solid basis for previous theories but also has contributed to further theoretical developments.

It is still quite surprising that, although actual interactions of filamentous bacteria are far more complicated than those in the Vicsek-style models and we do not know governing equations for all the bacterial activities, our system can be reduced to the Vicsek-style models and the corresponding hydrodynamic theories. Interactions of bacteria are basically dominated by excluded volume interaction, which is actually very difficult to model accurately in the numerical studies. Short-range (near-field) hydrodynamic interactions might also be required to model our experimental systems precisely because our filamentous bacteria are still swimming in such a thin layer of fluid by rapidly rotating their multiple flagella.

In comparison with numerical studies, the number of such 'microswimmers' with excluded volume that numerical simulations are capable of handling well depends on the modeling level, and such simulations are usually challenging. For instance, $\sim 10^6$ size-less point-like particles with simple interactions (as in the Vicsek-style models) are easily handled numerically. However, simulations of actual rod-like, or even spherical, swimmers with hydrodynamic interactions can hardly deal with $\sim 10^4$ elements [21–23, 49]. In our experiments, the number of bacteria in the observed area is 10^3–10^4, and $> 10^6$ in the chamber. This number is thus comparable, if not overwhelming, to that reachable in simulations of swimmers. For the moment, there is no numerical study with excluded volume and/or near-field hydrodynamics that is reduced to the Vicsek class. We hope that future numerical studies will confirm what is going on inside quasi-2D microswimmer systems like ours.

Our findings, like those of the Vicsek-style self-propelled rods [5], challenge existing theoretical works on the original Vicsek model and active nematics. The legitimate question we introduced in Sect. 2.4.2, raised in past works [12], is whether self-propelled 'rods' constitute an entirely different class from polar flocks and active nematics. Due to the symmetry of the ordered phases, some speculate without any solid theoretical basis that self-propelled rods should be the same as active nematics. However, our experimental results, especially the existence of true long-range order, do support that the self-propelled rods are different from active nematics but exhibit similar phenomenology as the original Vicsek model.

Our experimental results also challenge the validity of hydrodynamic theories and can contribute further theoretical development. Because there are still no existing analytical results on the Vicsek-style self-propelled rods, we compared our experimentally obtained exponents of GNF and the correlation functions with those in the Toner-Tu theory for the original Vicsek model. Some exponents coincide, but the others do not. This raises many questions and possibilities: Are the exponents for self-propelled rods different from those for the Vicsek model? If not, does the forgotten anisotropic pressure term P_2 modify the first theoretical predictions? Do some factors that are present in actual experiments such as excluded volume alter the exponents, although the Toner-Tu predictions are correct?

Though such controversy should still be resolved by rigorous theoretical calculations, our results provide the first unambiguous, large-scale, experimental evidence of the characteristic properties of order and fluctuations in globally-ordered homogeneous active phases predicted by the standard models of aligning self-propelled particles. In this context, future work will focus on obtaining better control on the density of bacteria so as to be able to study the transition to nematic order. At the biological level, one could speculate that the long-range correlations put forward here might provide a means to collectively probe scales far beyond the individual cell's capacity.

3.6 Appendix

3.6.1 Structure Tensor Method and Coherency

Here we summarize the structure tensor method and the coherency C [33, 50]. The structure tensor method is used for detecting the 'orientation' of images.[9] Here we

---

[9]In the field of image analysis, the term 'orientation' is often used to represent the direction of images in which periodic patterns appear. The 'orientation' in this definition is different by $90°$ from the direction of nematic director field that we want to know.

define the 'orientation' of images as the direction with the minimal variation, in accordance with the direction of nematic director fields that we want to measure. In the grayscale images such as fluorescent images, it corresponds to finding the direction with the smallest intensity gradient.

Suppose we have an image data $f(x, y)$ in a region of interest. In order to find the orientation of $f(x, y)$, then we consider the derivative in a certain direction specified by a unit vector $\boldsymbol{u}_\theta$ directing in the θ-direction,

$$\boldsymbol{u}_\theta^T \nabla f(x, y), \tag{3.40}$$

where,

$$\boldsymbol{u}_\theta := \begin{pmatrix} \cos\theta \\ \sin\theta \end{pmatrix}, \tag{3.41}$$

and the superscript T means transposition. Then, we look for the direction that gives the smallest norm of $\boldsymbol{u}_\theta^T \nabla f(x, y)$. If we define the inner product as,

$$\langle g, h \rangle = \iint_{\text{ROI}} g\, h\, \mathrm{d}x\, \mathrm{d}y, \tag{3.42}$$

then the squared norm is given by,

$$\|\boldsymbol{u}_\theta^T \nabla f(x, y)\|^2 = \langle \boldsymbol{u}_\theta^T \nabla f(x, y), \nabla f(x, y)^T \boldsymbol{u}_\theta \rangle = \boldsymbol{u}_\theta^T \boldsymbol{J} \boldsymbol{u}_\theta, \tag{3.43}$$

where $\boldsymbol{J}$ is the structure tensor,

$$\boldsymbol{J} = \begin{bmatrix} \langle f_x, f_x \rangle & \langle f_x, f_y \rangle \\ \langle f_x, f_y \rangle & \langle f_y, f_y \rangle \end{bmatrix}. \tag{3.44}$$

Then, the problem of finding the directions of minimal/maximal $\|\boldsymbol{u}_\theta^T \nabla f(x, y)\|^2$ is reduced to the eigenvalue problem of the 2×2 symmetric positive-definite matrix $\boldsymbol{J}$.

By explicitly solving this problem, we can obtain the direction θ that gives the maximum of the intensity gradient $\|\boldsymbol{u}_\theta^T \nabla f(x, y)\|^2$, which is different from the orientation in our definition by $90°$, as,

$$\theta = \frac{1}{2} \arctan\left(\frac{2\langle f_x, f_y \rangle}{\langle f_y, f_y \rangle - \langle f_x, f_x \rangle} \right). \tag{3.45}$$

The coherency C is defined as,

$$C := \frac{\lambda_{\max} - \lambda_{\min}}{\lambda_{\max} + \lambda_{\min}} = \frac{\sqrt{(\langle f_y, f_y \rangle - \langle f_x, f_x \rangle)^2 + 4\langle f_x, f_y \rangle^2}}{\langle f_x, f_x \rangle + \langle f_y, f_y \rangle}. \tag{3.46}$$

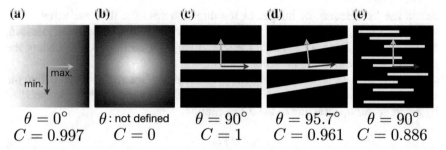

Fig. 3.20 The direction θ and the coherency C are calculated for 5 test images. The two eigenvectors for each image are depicted. The smallest eigenvector of $\mathbf{J}$ depicted as blue arrows gives the orientation of images that correspond to the nematic director field as we can see from (**c**). $C = 0$ for isotropic images like (**b**), and $C \simeq 1$ for strongly anisotropic images. Because the structure tensor detects the gradients, even the images like (**e**) cannot give $C = 1$, although such images have the scalar nematic order parameter $Q = 1$

Note that the coherency C takes the value between 0 and 1, $C \in [0, 1]$. This coherency C is a measure of anisotropy of images.

The coherency C and the θ calculated for some test images are shown in Fig. 3.20 in order to grasp the intuitive idea of them. Although C can be used as an estimate of the nematic order parameter, there is no analytical relation because this structure tensor analysis depends on shapes of each object inside the images and even the complete nematic alignment cannot sometimes give $C = 1$ as in Fig. 3.20e.

3.6.2   Calculation of Correlation Function of Director Fluctuations $C_{\delta n_\perp}(\mathbf{R})$

The correlation function $C_{\delta n_\perp}(\mathbf{R})$ was calculated from the director field obtained by the structure tensor method. We took ROIs of 64×64 pixels shifted by 8 pixels in either direction (87.5 % overlap, 249×249 boxes in total). We calculated the local direction $\theta(\mathbf{r}, t)$ of the director field for each 64×64-pixel ROI from the structure tensor method [33]. Using $\theta(\mathbf{r}, t)$, we can calculate a correlation function,

$$\langle \delta n_\perp(\mathbf{r}, t) \delta n_\perp(\mathbf{r} + \mathbf{R}, t) \rangle_r \,, \tag{3.47}$$

which is used for nematic liquid crystals [36]. As is often the case, we assume the deviation $\delta n_\perp = n - n_0$ of the local director n from the global mean director n_0 is small. Because $|n|^2 = |n_0|^2 = 1$ holds by definition,

$$|n|^2 = |n_0|^2 + n_0 \cdot \delta n_\perp + |\delta n_\perp|^2 \tag{3.48}$$
$$\simeq 1 + n_0 \cdot \delta n_\perp \,, \tag{3.49}$$

and hence $n_0 \cdot \delta n_\perp = 0$ holds, meaning the fluctuations of the director field have only transverse components. The transverse component $\delta n_\perp$ of $\delta n_\perp$ can be obtained as,

$$\delta n_\perp(r, t) = \sin\left[\theta(r, t) - \frac{1}{A}\int\theta(r, t)d^2r\right], \tag{3.50}$$

where the integral is over the whole field of view and A is its area. Then we calculated $\langle\delta n_\perp(r, t)\delta n_\perp(r + R, t)\rangle_r$ and then averaged over time t,

$$C_{\delta n_\perp}(R) := \langle\langle\delta n_\perp(r, t)\delta n_\perp(r + R, t)\rangle_r\rangle_t. \tag{3.51}$$

To reduce the computational cost and to avoid calculating correlated successive images, we used every 20 frames of the movie (100 frames in total).

3.6.3 Relations Among Nematic Order Parameters: $|\langle e^{2i\theta}\rangle|, S, \boldsymbol{S}, \boldsymbol{Q}, \hat{f}_2$

Here we summarize the relations among 4 nematic order parameters: a frequently used parameter $|\langle e^{2i\theta}\rangle|$, the scalar nematic order S, the tensorial nematic order $\boldsymbol{Q}$, and the 2nd order Fourier mode $\hat{f}_2(r, t)$ of the one-particle distribution function $f(r, \theta, t)$.

Definition

When $\boldsymbol{a}$ denotes the unit vector representing the orientation of particles such as liquid crystal molecules and self-propelled particles, the usual definitions of nematic order parameters in a d-dimensional space are given as,

$$S_{ij} = \frac{d}{d-1}(\langle a_i a_j\rangle - \frac{1}{d}\delta_{ij}), \tag{3.52}$$

$$Q_{ij} = \langle a_i a_j\rangle - \frac{1}{d}\delta_{ij}, \tag{3.53}$$

where $\langle\ \rangle$ is a local average, and i and j denote spatial coordinates. The above S_{ij} is normalized so that its largest eigenvalue takes the maximum value 1 when all the particles are aligned exactly in the same direction. The largest eigenvalue of S_{ij} is called 'the scalar order parameter' and often denoted as S.

On the other hand, Q_{ij} is not normalized and its eigenvalue in the complete alignment is $\frac{d-1}{d}$. In the present-day studies, we often denote the above S_{ij} as Q_{ij}. However, in the studies on kinetic theories of collective motion [39, 40, 51], Q_{ij} defined in Eq. (3.53) above is frequently used due to a useful relation with the Fourier mode as we describe below.

Hereafter, we will consider only the case of $d = 2$.

Q_{ij} and $\hat{f}_2(\boldsymbol{r}, t)$

First, let us clarify the relation between Q_{ij} and $\hat{f}_2(\boldsymbol{r}, t)$. Because Q_{ij} is a traceless symmetric tensor, the components of Q_{ij} can be simplified as,

$$Q_{ij} = \langle a_i a_j \rangle - \frac{1}{2}\delta_{ij} \tag{3.54}$$

$$= \begin{pmatrix} Q_{11} & Q_{12} \\ Q_{21} & Q_{22} \end{pmatrix} \tag{3.55}$$

$$= \begin{pmatrix} Q_{11} & Q_{12} \\ Q_{12} & -Q_{11} \end{pmatrix}. \tag{3.56}$$

Here, if we denote the one-particle distribution function as $f(\boldsymbol{r}, \theta, t)$ with θ measured with respect to the x-axis, then the local density field $\rho(\boldsymbol{r}, t)$ is given by,

$$\rho = \int_{-\pi}^{\pi} d\theta\, f(\boldsymbol{r}, \theta, t) = \int_{-\pi}^{\pi} d\theta\, f(\boldsymbol{r}, \theta, t)e^{0i\theta} = \hat{f}_0(\boldsymbol{r}, t). \tag{3.57}$$

Therefore, the 0th order Fourier mode $\hat{f}_0(\boldsymbol{r}, t)$ is nothing but the local density field $\rho(\boldsymbol{r}, t)$, and according to this fact, we have to be careful about how the distribution function is normalized: The distribution function is *not* a simple probability distribution function whose integrated value equals 1. With this in mind, we can rewrite the components of Q_{ij} as,

$$Q_{11} = \langle a_1{}^2 \rangle - \frac{1}{2}\delta_{11} \tag{3.58}$$

$$= \langle \cos^2 \theta \rangle - \frac{1}{2} \tag{3.59}$$

$$= \left\langle \frac{\cos 2\theta + 1}{2} \right\rangle - \frac{1}{2} \tag{3.60}$$

$$= \frac{1}{2} \langle \cos 2\theta \rangle \tag{3.61}$$

$$= \frac{1}{2}\mathrm{Re} \int_{-\pi}^{\pi} d\theta\, \frac{f(\boldsymbol{r}, \theta, t)}{\rho} e^{2i\theta}, \tag{3.62}$$

$$Q_{12} = \langle a_1 a_2 \rangle - \frac{1}{2}\delta_{12} \tag{3.63}$$

$$= \langle \cos \theta \sin \theta \rangle \tag{3.64}$$

$$= \left\langle \frac{1}{2} \sin 2\theta \right\rangle \tag{3.65}$$

$$= \frac{1}{2}\mathrm{Im} \int_{-\pi}^{\pi} d\theta\, \frac{f(\boldsymbol{r}, \theta, t)}{\rho} e^{2i\theta}. \tag{3.66}$$

Therefore, by using the 2nd order Fourier mode $\hat{f}_2(\boldsymbol{r}, t)$ defined as,

$$\hat{f}_2 = \int_{-\pi}^{\pi} d\theta \, f(\boldsymbol{r}, \theta, t) e^{2i\theta}, \tag{3.67}$$

we can represent Q_{ij} in terms of $\hat{f}_2$ as,

$$\rho Q_{ij} = \frac{1}{2} \begin{pmatrix} \operatorname{Re}\hat{f}_2 & \operatorname{Im}\hat{f}_2 \\ \operatorname{Im}\hat{f}_2 & -\operatorname{Re}\hat{f}_2 \end{pmatrix}. \tag{3.68}$$

Such relations are very useful for analyzing the Boltzmann equation and obtaining hydrodynamic equations from that.

$|\langle e^{2i\theta} \rangle|$, S, and Q_{ij}

In some experiments [32], $|\langle e^{2i\theta} \rangle|$ is often used as an estimate of nematic order. This can be rewritten as,

$$|\langle e^{2i\theta} \rangle| = |\langle \cos 2\theta \rangle + i \langle \sin 2\theta \rangle| \tag{3.69}$$

$$= 2|Q_{11} + i Q_{12}| \tag{3.70}$$

$$= 2\sqrt{Q_{11}^2 + Q_{12}^2}. \tag{3.71}$$

On the other hand, eigenvalues λ of Q_{ij} are easily calculated from the characteristic equation of Q_{ij} as,

$$\lambda^2 - (Q_{11}^2 + Q_{12}^2) = 0 \tag{3.72}$$

$$\Leftrightarrow \lambda = \pm\sqrt{Q_{11}^2 + Q_{12}^2}. \tag{3.73}$$

Therefore, $|\langle e^{2i\theta} \rangle|$ is the same as the largest eigenvalue of the normalized tensorial order parameter $S_{ij} = 2Q_{ij}$, i.e. the scalar order parameter S.

References

1. Vicsek T, Czirók A, Ben-Jacob E, Cohen I, Shochet O (1995) Novel type of phase transition in a system of self-driven particles. Phys Rev Lett 75(6):1226
2. Grégoire G, Chaté H (2004) Onset of collective and cohesive motion. Phys Rev Lett 92(2):025702
3. Chaté H, Ginelli F, Montagne R (2006) Simple model for active nematics: quasi-long-range order and giant fluctuations. Phys Rev Lett 96(18):180602
4. Chaté H, Ginelli F, Grégoire G, Peruani F, Raynaud F (2008) Modeling collective motion: variations on the vicsek model. Eur Phys J B 64:451–456
5. Ginelli F, Peruani F, Bär M, Chaté H (2010) Large-scale collective properties of self-propelled rods. Phys Rev Lett 104(18):184502

6. Ngo S, Peshkov A, Aranson IS, Bertin E, Ginelli F, Chaté H (2014) Large-scale chaos and fluctuations in active nematics. Phys Rev Lett 113(3):038302
7. Toner J, Tu Y (1995) Long-range order in a two-dimensional dynamical XY model: how birds fly together. Phys Rev Lett 75(23):4326–4329
8. Toner J, Tu Y (1998) Flocks, herds, and schools: a quantitative theory of flocking. Phys Rev E 58(4):4828–4858
9. Toner J, Tu Y, Ramaswamy S (2005) Hydrodynamics and phases of flocks. Ann Phys 318:170–244
10. Toner J (2012) Reanalysis of the hydrodynamic theory of fluid, polar-ordered flocks. Phys Rev E 86(3):031918
11. Ramaswamy S, Simha RA, Toner J (2003) Active nematics on a substrate: Giant number fluctuations and long-time tails. Europhys Lett 62(2):196–202
12. Marchetti MC, Joanny JF, Ramaswamy S, Liverpool TB, Prost J, Rao M, Simha RA (2013) Hydrodynamics of soft active matter. Rev Modern Phys 85(3):1143–1189
13. Peruani F, Starruß J, Jakovljevic V, Søgaard-Andersen L, Deutsch A, Bär M (2012) Collective motion and nonequilibrium cluster formation in colonies of gliding bacteria. Phys Rev Lett 108(9):098102
14. Zhang HP, Be'er A, Florin E-L, Swinney HL (2010) Collective motion and density fluctuations in bacterial colonies. Proc Natl Acad Sci USA, Vol 107, No 31, pp 13626–13630
15. Sokolov A, Aranson IS, Kessler JO, Goldstein R (2007) Concentration dependence of the collective dynamics of swimming bacteria. Phys Rev Lett 98(15):158102
16. Sokolov A, Aranson IS (2012) Physical properties of collective motion in suspensions of bacteria. Phys Rev Lett 109(24):248109
17. Wensink HH, Dunkel J, Heidenreich S, Drescher K, Goldstein RE, Löwen H, Yeomans JM (2012) Meso-scale turbulence in living fluids. Proc Natl Acad Sci USA 109(36):14308–14313
18. Dunkel J, Heidenreich S, Drescher K, Wensink HH, Bär M, Goldstein RE (2013) Fluid dynamics of bacterial turbulence. Phys Rev Lett 110(22):228102
19. Gachelin J, Rousselet A, Lindner A, Clement E (2014) Collective motion in an active suspension of Escherichia coli bacteria. New J Phys 16(2):025003
20. Subramanian G, Koch DL (2009) Critical bacterial concentration for the onset of collective swimming. J Fluid Mech 632:359
21. Saintillan D, Shelley MJ (2007) Orientational order and instabilities in suspensions of self-locomoting rods. Phys Rev Lett 99(5):058102
22. Saintillan D, Shelley MJ (2008) Instabilities and pattern formation in active particle suspensions: Kinetic theory and continuum simulations. Phys Rev Lett 100(17):178103
23. Saintillan D, Shelley MJ (2012) Emergence of coherent structures and large-scale flows in motile suspensions. J Royal Soc Interface Royal Soc 9(68):571–85
24. Drescher K, Dunkel J, Cisneros LH, Ganguly S, Goldstein RE (2011) Fluid dynamics and noise in bacterial cell–cell and cell–surface scattering. Proc Natl Acad Sci USA 108:10940–10945
25. Lefauve A, Saintillan D (2014) Globally aligned states and hydrodynamic traffic jams in confined suspensions of active asymmetric particles. Phys Rev E 89(2):021002(R)
26. Wensink HH, Löwen H (2012) Emergent states in dense systems of active rods: from swarming to turbulence. J Phys: Condensed Matter Condensed Matter 24(46):464130
27. Takeuchi S, Diluzio WR, Weibel DB, Whitesides GM (2005) Controlling the shape of filamentous cells of Escherichia coli. Nano Lett 5(9):1819–1823
28. Maki N, Gestwicki JE, Lake EM, Kiessling LL, Adler J (2000) Motility and chemotaxis of filamentous cells of escherichia coli. J Bacteriol 182(15):4337–4342
29. Scharf BE, Fahrner KA, Turner L, Berg HC (1998) Control of direction of flagellar rotation in bacterial chemotaxis. Proc Natl Acad Sci USA 95(1):201–206
30. Lauga E, DiLuzio WR, Whitesides GM, Stone HA (2006) Swimming in circles: motion of bacteria near solid boundaries. Biophys J 90(2):400–412
31. Swiecicki J-M, Sliusarenko O, Weibel DB (2013) From swimming to swarming: escherichia coli cell motility in two-dimensions. Integr Biol: Quant Biosci Nano to Macro 5:1490–1494

32. Nishiguchi D, Sano M (2015) Mesoscopic turbulence and local order in Janus particles self-propelling under an ac electric field. Phys Rev E 92(5):052309
33. Rezakhaniha R, Agianniotis A, Schrauwen JTC, Griffa A, Sage D, Bouten CVC, van de Vosse FN, Unser M, Stergiopulos N (2012) Experimental investigation of collagen waviness and orientation in the arterial adventitia using confocal laser scanning microscopy. Biomech Model Mechanobiol 11:461–473
34. Otsu N (1979) A threshold selection method from gray-level histograms. IEEE Trans Syst Man Cybern 9(1):62–66
35. Chaté H, Ginelli F, Grégoire G, Raynaud F (2008) Collective motion of self-propelled particles interacting without cohesion. Phys Rev E 77(4):046113
36. Landau LD, Lifshitz EM (1980) Statistical physics. Elsevier, 3rd edn
37. Suzuki R, Weber CA, Frey E, Bausch AR (2015) Polar pattern formation in driven filament systems requires non-binary particle collisions. Nature Phys 11:839–844
38. Hanke T, Weber CA, Frey E (2013) Understanding collective dynamics of soft active colloids by binary scattering. Phys Rev E 88(5):052309
39. Peshkov A, Aranson IS, Bertin E, Chaté H, Ginelli F (2012) Nonlinear field equations for aligning self-propelled rods. Phys Rev Lett 109(26):268701
40. Peshkov A, Bertin E, Ginelli F, Chaté H (2014) Boltzmann-Ginzburg-Landau approach for continuous descriptions of generic Vicsek-like models. Eur Phys J Special Topics 223:1315–1344
41. Weber CA, Thüroff F, Frey E (2013) Role of particle conservation in self-propelled particle systems. New J Phys 15:045014
42. Duclos G, Garcia S, Yevick HG, Silberzan P (2014) Perfect nematic order in confined monolayers of spindle-shaped cells. Soft Matter 10(14):2346–2353
43. Narayan V, Ramaswamy S, Menon N (2007) Long-lived giant number fluctuations in a swarming granular nematic. Science (New York, N.Y.) 317:105
44. Deseigne J, Dauchot O, Chaté H (2010) Collective motion of vibrated polar disks. Phys Rev Lett 105(9):098001
45. Kumar N, Soni H, Ramaswamy S, Sood AK (2014) Flocking at a distance in active granular matter. Nature Commun 5:4688
46. Shi X-q, Chaté H (2018) Self-propelled rods: linking alignment-dominated and repulsion-dominated active matter. arXiv: 1807.00294
47. Tanida S, Furuta K, Nishikawa K, Hiraiwa T, Kojima H, Oiwa K, Sano M (2018) Gliding filament system giving both orientational order and clusters in collective motion. arXiv: 1806.01049
48. Maitra A, Srivastava P, Marchetti MC, Lintuvuori JS, Ramaswamy S, Lenz M (2018) A nonequilibrium force can stabilize 2D active nematics. Proc Natl Acad Sci USA 115(27):6934–6939
49. Ishikawa T, Pedley TJ (2008) Coherent structures in monolayers of swimming particles. Phys Rev Lett 100(8):088103
50. Jähne B (1993) Spatio-temporal image processing: theory and scientific applications, 1st edn. Springer, Berlin, Heidelberg
51. Bertin E, Chaté H, Ginelli F, Mishra S, Peshkov A, Ramaswamy S (2013) Mesoscopic theory for fluctuating active nematics. New J Phys 15(8):085032

Chapter 4
Active Turbulence

Abstract In this chapter, we review both theoretical and experimental studies on active turbulence, especially bacterial turbulence. Hydrodynamic flows created by bacteria usually destabilize alignment, leading to turbulent states without any long-range order. Such turbulent states resemble fluid turbulence in spite of their low Reynolds number. Hydrodynamic theories have succeeded in describing the properties of bulk unconstrained bacterial turbulence, but the boundary conditions of active turbulence remain unknown. Recent experiments have found that bacterial turbulence geometrically confined in droplets or in microfluidic devices can self-organize in vortex states, which suggests the importance of boundary conditions to understand macroscopic behavior of bacterial turbulence. At the end of this chapter, we summarize what needs to be experimentally investigated for a better understanding.

Keywords Bacterial turbulence · Hydrodynamic instability
Boundary conditions · Vortex order · Janus particle

4.1 Bacterial Turbulence

In Chaps. 2 and 3, we have investigated collective motion that belongs to the Vicsek universality class. However, such collective motion has never been found before our experiment on filamentous bacteria in quasi-two-dimensions. Instead, what we frequently observe is turbulent collective motion. Such turbulent motion can be observed in dense bacterial suspensions [1–5], sperms [6], self-propelling colloids [7], and even disastrous crowd dynamics of people [8]. Especially, due to experimental accessibility, bacterial turbulence is well investigated.

Such turbulent states, especially those seen in microswimmer suspensions, are now termed 'active turbulence', 'mesoscopic turbulence', or 'mesoscale turbulence'. This is another class of collective motion that is also well investigated both theoretically and experimentally.

Unlike the Vicsek-style models, hydrodynamics plays an important role in dynamics of bacterial turbulence. Steric interactions of bacteria usually work to align bacteria as we have seen in Chap. 3, but hydrodynamic flow created by bacteria usually

© Springer Nature Singapore Pte Ltd. 2020
D. Nishiguchi, *Order and Fluctuations in Collective Dynamics
of Swimming Bacteria*, Springer Theses, https://doi.org/10.1007/978-981-32-9998-6_4

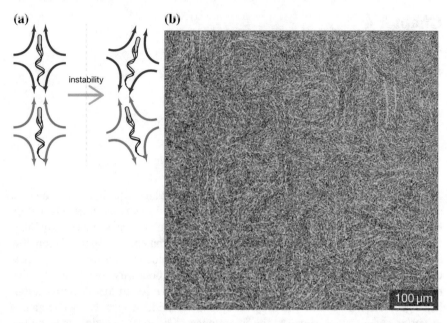

Fig. 4.1 **a** Schematics of hydrodynamic instability arising from bacterial pusher-type flow. **b** Snapshot of bacterial turbulence overlaid with instantaneous velocity field. The velocity field is calculated from particle image velocimetry (PIV)

destabilizes such alignment, leading to turbulent states without any long-range order (Fig. 4.1). Hydrodynamics triggers not only such instability but also interesting collective acceleration. Characteristic speed of emergent flows in active turbulence is surprisingly 3–5 times faster than that of an isolated single bacterium. Therefore, active turbulence is genuinely a collective phenomenon.

Although we call such states 'active turbulence', of course, active turbulence is distinctively different from classical fluid turbulence as seen in water. Turbulent flow in a fluid is usually associated with high Reynolds numbers ($Re > 5 \times 10^4$), at which inertia dominates viscosity. On the other hand, swimming bacteria, such as *Bacillus subtilis* or *Escherichia coli*, are living in extremely low Reynolds number. Because they are usually only ~5 μm long and swim at the speed of ≈20 μm/s, the Reynolds number for fluid flow created by a single bacterium is of the order of $10^{-5} - 10^{-4}$ [9]. However, at high concentration of bacteria, their dynamics resembles fluid turbulence in that their macroscopic velocity field continuously changes and there exist many jets, flows, and vortices [1, 2, 4, 5]. In this sense, it is a good idea to investigate active turbulence in analogy with fluid turbulence.

4.2 Basic Properties of Bacterial Turbulence

4.2.1 Constant Correlation Length

In bacterial turbulence, we can observe collective motion of bacteria over the length scale larger than individual bacteria. As a simple measure of collective motion, we can define a correlation function of the velocity field of bacterial turbulence which can be estimated by particle image velocimetry (PIV) programs (Fig. 4.1b).

Andrey Sokolov et al. conducted a series of well-controlled experiments on the measurements of the correlation length of bacterial turbulence [4, 10]. They first controlled the density of bacteria *Bacillus subtilis* in a suspension by utilizing their chemotactic response [4]. By applying an electric current, they tuned pH of suspensions locally. This allowed them to vary the density in a single trial of experiments. As a result, above threshold density for the onset of turbulent states, they obtained a constant correlation length. Even though the collective swimming speed changes as increasing the bacterial density, the correlation length stays constant (Fig. 4.2a).

They also varied individual swimming speed by tuning the oxygen concentration of surrounding air [10]. Because *Bacillus subtilis* is an aerobic bacterium, its swimming speed strongly depends on oxygen concentration. They constructed a chamber that can control the concentration of oxygen inside, and they succeeded in repeatedly increasing and decreasing the individual swimming speed of bacteria. Using this

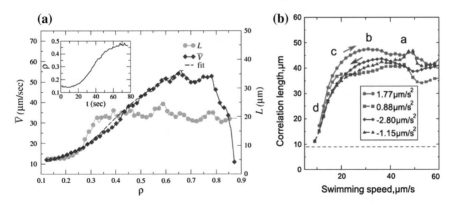

Fig. 4.2 **a** Correlation length L and collective swimming speed $\overline{V}$ versus the filling fraction of bacteria ρ. Correlation length stays constant above threshold density. **b** Correlation length versus individual swimming speed. By changing the concentration of oxygen in the air, they repeatedly varied the swimming speed of bacteria. Correlation length is again almost constant above threshold swimming speed. Figure **a** and **b** reproduced from [4] and [10] respectively

setup, they measured the correlation length as a function of the individual swimming speed. Surprisingly, the correlation length was again almost constant above threshold swimming speed for turbulent states (Fig. 4.2b). This indicates that the individual swimming speed just affects the time scale of bacterial turbulence and does not alter other properties.

Such constant correlation lengths can be understood as a result of competition between steric interactions and hydrodynamic interactions [10]. Steric interactions, or collisions, alone work to align bacteria, as we have seen in our experiments on filamentous bacteria in Chap. 3 [11]. The expansion rate of such aligned regions can be estimated from the mean free time t_0 of bacteria between collisions in 3-dimensions,

$$t_0 = \frac{1}{\rho l^2 V}, \tag{4.1}$$

where ρ is the number density of bacteria, l is the length of bacteria, and V is the velocity of bacteria. Therefore, if we assume that every collision increases the linear size of the region by l, the expansion speed of the aligned regions is estimated as,

$$V_c \sim \frac{l}{t_0} = \rho l^3 V. \tag{4.2}$$

On the other hand, hydrodynamic interactions destabilize such alignment (Fig. 4.1). Hydrodynamic flow created by a single bacterium, or a single force dipole, swimming in the x-direction can be written as,

$$u(r) = \frac{pr}{8\pi r^3} \left(\frac{3x^2}{r^2} - 1 \right) \propto \frac{p}{r^2}, \tag{4.3}$$

where $p = aVl^2$ is the strength of the force dipole with a geometrical constant a. The flow V_h created by the bacteria inside the aligned region at the distance R can be estimated as,

$$V_h \sim \rho \cdot \frac{4}{3}\pi R^3 \cdot \frac{p}{R^2} \sim \rho p R. \tag{4.4}$$

Then, we consider the distance R at which both the collision-induced expansion speed of the aligned region and the strength of destabilizing flow field are balanced, and we obtain,

$$V_c \sim V_h \tag{4.5}$$

$$\Rightarrow R \sim \frac{l^3 V}{p} = \frac{l}{a} \simeq 5 - 10l, \tag{4.6}$$

which gives a value consistent with experiments. Here, the experimentally obtained value for a was inserted [12]. Therefore, the density ρ and the swimming speed V are canceled out and do not appear in the final representation Eq. (4.6).

The above two experimental facts mean that bacterial turbulence is a very nice system with a constant correlation length irrespective of the density and the individual swimming speed. These assure that we can almost always reproduce the same result from other trials of experiments. We do not need to care so much about the reproducibility of experiments on bacterial turbulence.

4.2.2 Power Spectrum

In analogy with classical fluid turbulence, power spectra of velocity fields of bacterial turbulence and active turbulence of self-propelling colloids were investigated [5, 7].

According to Kolmogorov's law [13], the power spectrum of usual fluid turbulence exhibits algebraic decay with a characteristic negative exponent. This algebraic behavior comes from the existence of conserved quantities. In usual fluid turbulence, energy is injected externally by boundary conditions such as pressure gradient and shear stress. This injected energy is then transferred to smaller scales and finally dissipates into heat at small eddies. Such energy cascade in the intermediate length scale is the origin of the power law.[1]

A crucial difference between classical fluid turbulence and active turbulence is in the spatial scales where energy is injected into the systems. In usual fluid turbulence, energy injection from the boundary occurs at the scale of its system size, i.e. the largest length scale. In contrast, in the active turbulence of the Janus particles [7] or bacteria [5], energy is injected at the particle level.

For active turbulence, characteristic power spectra were obtained in bacterial turbulence (Fig. 4.3) [5]. They have a peak at around ~10 bacterial body length, which corresponds to the correlation length of bacterial turbulence. Beside the peak, algebraic behavior was observed for both higher and lower wavenumber regimes, although the observed range was quite limited. Such behavior was reproduced by simulations on self-propelled rods with excluded volume [5, 14] and a continuum theory [5]. However, later studies on an agent-based model [15] that exhibits active turbulence and on the continuum theory [16] have proved that the exponents of the power law in the power spectrum of active turbulence are not universal and do depend on the values of parameters. In accordance with these theoretical studies, we found that, in our previous experiments, turbulent states of self-propelling asymmetrical colloidal particles (Janus particles) under an AC electric field also exhibit a similar power spectrum shape but with different exponents (Fig. 4.4).

---

[1] In addition to energy conservation, enstrophy is also conserved in two-dimensional fluid turbulence. This additional conservation law modifies the exponent in two-dimensional turbulence from that in three-dimensional turbulence.

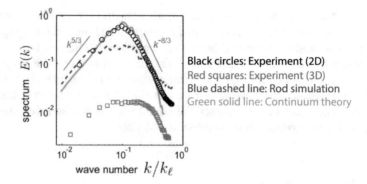

Fig. 4.3 The power spectrum of velocity fields of bacterial turbulence shows a peak, which testifies the existence of characteristic length scale of ∼10 bacterial body length (correlation length). The wavenumber is normalized by the length of bacteria. The power law behavior was suggested in [5], but later studies on an agent-based model and continuum theories indicate the exponents of the power law are non-universal. Figure modified and reproduced from [5]

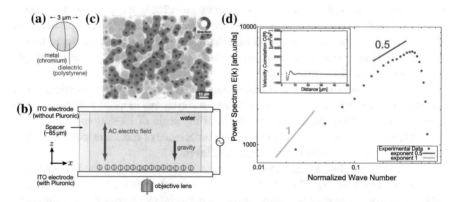

Fig. 4.4 Our previous experiment on asymmetrical colloidal particles (Janus particles) fueled by an AC electric field [7, 17]. **a** Schematics of our particles. Half hemispheres of the dielectric colloids are coated with metal. **b** Experimental setup. A vertical electric field was applied to the Janus particles by using two transparent planar electrodes, and consequently they propel in a horizontal plane close to the bottom electrode. **c** Colors indicating directions of motion were overlaid on an experimental snapshot. There are clusters with characteristic lengths of ∼4 particle diameters. **d** The power spectrum of the velocity field of the Janus particles has a similar shape but different exponents. Inset: velocity correlation function. Solid lines are just to guide the eye. Figures reproduced from [7]

4.3 Continuum Theory

4.3.1 Formalism

To describe active turbulence, continuum hydrodynamic theories have been devised. Most of them include many parameters and many variables such as local density fields, velocity fields, tensorial nematic order parameters, etc. [18] and hence they are quite difficult to test experimentally. In need of tractable models comparable with experiments, a minimal model with a few variables was devised [3, 5, 19].[2] Using this model, the power spectrum of the velocity field was reproduced (Green line in Fig. 4.3).

First, because we are thinking about very dense suspensions, it is a good approximation to postulate incompressibility,

$$\nabla \cdot \boldsymbol{v} = \partial_i v_i = 0, \tag{4.7}$$

where $\boldsymbol{v}$ is the velocity field of bacteria, and v_i is the i-component of $\boldsymbol{v}$. Then, the generalized Navier-Stokes equation is assumed,

$$(\partial_t + \boldsymbol{v} \cdot \nabla)\boldsymbol{v} = -\nabla p - (A + C|\boldsymbol{v}|^2)\boldsymbol{v} + \nabla \cdot \boldsymbol{E}, \tag{4.8}$$

where the pressure $p(\boldsymbol{r}, t)$ is the Lagrange multiplier for the incompressibility constraint. The second term on the right-hand side, $(A + C|\boldsymbol{v}|^2)\boldsymbol{v}$, is similar to one in the Toner-Tu equations in Eq. (2.9) that again comes from a Ginzburg-Landau type potential $\frac{A}{2}|\boldsymbol{v}|^2 + \frac{C}{4}|\boldsymbol{v}|^4$. This term represents self-propulsion of bacteria, meaning non-zero $\boldsymbol{v}$. The last term is the most important for the emergence of active turbulence. As we have schematically explained in Fig. 4.1a, hydrodynamic instability that bends the orientation field of bacteria has to be included in the equation. In this spirit, the symmetric traceless strain rate tensor $\boldsymbol{E}$ is given by,

$$E_{ij} = \Gamma_0(\partial_i v_j + \partial_j v_i) - \Gamma_2 \Delta(\partial_i v_j + \partial_j v_i) + S q_{ij}, \tag{4.9}$$

where S is a constant and q_{ij} is the tensorial nematic order parameter. The first Γ_0 term is required for the equation to contain the Navier-Stokes equations as a limiting case. The Γ_2 term is introduced so that the equations to be damped at higher wavenumbers. To further reduce the variables, q_{ij} is approximated as,

$$q_{ij} = v_i v_j - \frac{\delta_{ij}}{d}|\boldsymbol{v}|^2, \tag{4.10}$$

---

[2]Of course, such a model with a few variables might not precisely describe actual phenomena, but this can be a first step toward a full understanding of active turbulence.

where d is the spatial dimension. This is a mean-field approximation by assuming that the director field n, or the orientation field, of bacteria relaxes sufficiently fast so that n can be effectively replaced by v. As for the constant S, from the physical arguments, $S < 0$ for pusher-type swimmers like *Escherichia coli* and *Bacillus subtilis*, and $S > 0$ for puller-type swimmers[3] like algae *Chlamydomonas*.

By substituting Eqs. (4.9) and (4.10) into Eq. (4.8), we finally obtain the equation of motion for active turbulence,

$$(\partial_t + \lambda_0 v \cdot \nabla)v = -\nabla p + \lambda_1 \nabla v^2 - (A + C|v|^2)v + \Gamma_0 \Delta v - \Gamma_2 \Delta^2 v, \quad (4.11)$$

where $\lambda_0 := 1 - S$ and $\lambda_1 := -S/d$.

4.3.2 Difficulty in Boundary Conditions

Although this equations with experimentally extracted parameters successfully described the behavior of bulk unconstrained bacterial turbulence and reproduced the power spectrum shown in Fig. 4.3 [5], it remains unclear how to treat boundary conditions on bacterial turbulence. If we want to understand what happens when bacterial turbulence is in contact with some boundaries, we have to incorporate boundary conditions to solve these partial differential equations as we usually do for the Navier-Stokes equations. However, we do not know what the boundary conditions are for the bacterial velocity field v.[4] Boundary conditions for bacterial turbulence are far more complicated than those of usual fluid: Bacteria are attracted to walls due to hydrodynamic coupling and the density distribution gets inhomogeneous; bacteria may be trapped at the interfaces due to hydrodynamic coupling and surface tensions; individual bacteria exhibit circular trajectories near solid surfaces (Fig. 3.7) or at water-air interfaces [20, 21]; bacteria sometimes swim upstream with respect to the flow of ambient fluid along the walls [22].

Therefore, we still do not know in general how to implement boundary conditions into continuum theories. In this sense, bacterial turbulence in contact with some boundaries remains elusive. Experiments have been conducted to seek for the phenomena arising from the interplay between bacterial turbulence and boundaries.

---

[3]Puller-type microswimmers swim by pulling fluid in front of them, and consequently the fluid behind the body is drugged toward the swimmers. Therefore, a single puller-type microswimmer can be denoted as a force dipole directing inward.

[4]Note that this v is the velocity field only for bacteria, *not* for fluid. We neglect the dynamics of ambient fluid.

4.4 Bacterial Turbulence in Confinement

To understand the interplay between bacterial turbulence and boundaries, behavior of bacterial turbulence confined in small droplets or in microfluidic devices has been investigated. Furthermore, many wild bacteria usually live in geometrically constrained environments, such as soil, porous media, and host cells. Therefore, it is also of biological importance to investigate the dynamics of bacteria close to boundaries.

4.4.1 Spiral Formation in a Droplet

Wioland et al. conducted experiments on bacterial turbulence of *Bacillus subtilis* confined in droplets [23]. By pipetting dense bacterial suspensions in mineral oil and then sandwiching them between two coverslips, they made flattened droplets of bacterial turbulence inside mineral oil (Fig. 4.5a). The sizes of typical droplets were $h \sim 25\,\mu\text{m}$ in height and $d = 10 - 150\,\mu\text{m}$ in diameter. At around $d = 30 - 70\,\mu\text{m}$, they observed vortex formation with a spiral configuration of bacterial orientations (Fig. 4.5b,c).

The velocity fields of those vortices in droplets have counter-rotating structures. For example, in Fig. 4.5b, bacteria in the bulk (away from the boundary) are swimming counterclockwise, but bacteria close to the boundary are swimming clockwise. Bacteria at the boundary spontaneously start to swim along the wall, forming 'edge currents', and consequently they make counter-rotating fluid flow in the bulk. Then, bacteria in the bulk are driven by this flow and rotate in that direction. In fact, such behavior was reproduced by their numerical simulations on continuum equations [23] and agent-based model [24] that incorporate the fluid velocity field and the bac-

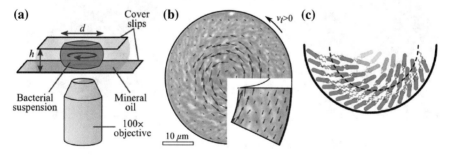

Fig. 4.5 **a** Schematic of the experimental setup in [23]. **b** Experimental snapshot with velocity field measured by PIV. Counterflow exists along the boundary. **c** Schematic explanation for the counterflow. Bacteria swimming along the wall in the clockwise direction make counterclockwise flow in the bulk, which then drives bacteria in the bulk. Thus, a counter-rotating spiral vortex is formed. Figures modified and reproduced from [23]

terial orientation field. Moreover, individual behavior of bacteria swimming along
the wall was experimentally confirmed by using fluorescent microscopy [24]. The
orientations of bacteria exhibit spiral configuration as shown in Fig. 4.5c.

4.4.2   Vortex Lattices in Connected Circular Cavities

As a next step, Wioland et al. investigated the dynamics of multiple vortices [25].
They fabricated microfluidic chambers with many circular cavities connected via
small channels (Fig. 4.6). Interestingly, the interaction through the small channels
gave rise to macroscopic 'ferromagnetic' or 'antiferromagnetic' order depending on
the channel gap width.

They quantified the vortices by defining the spin $V_i(t)$ of i-th cavity at time t as
the normalized planar angular momentum, and calculated the spin-spin correlation,

$$\chi = \left\langle \frac{\sum_{i\sim j} V_i(t)V_j(t)}{\sum_{i\sim j} |V_i(t)V_j(t)|} \right\rangle, \qquad (4.12)$$

where $\sum_{i\sim j}$ denotes a sum over all the pairs of adjacent cavities, and $\langle\ \rangle$ denotes
an average over all the experimentally obtained frames. When the gap width was
small, antiferromagnetic order $\chi < 0$ emerged. On the other hand, ferromagnetic
order $\chi > 0$ was observed for the large gap width case (Fig. 4.6).

The emergence of ferromagnetic and antiferromagnetic order was explained by the
competition between hydrodynamic continuity and edge currents induced by bacteria
swimming along the walls (Fig. 4.7). As we have seen in Sect. 4.4.1, bacteria at the
boundary tend to swim along the walls, forming 'edge currents' [23]. These edge

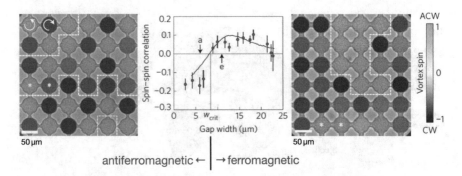

Fig. 4.6 Ferromagnetic (right) and antiferromagnetic (left) order of bacterial vortices in connected
microfluidic cavities. Small/large gap widths give rise to antiferromagnetic/ferromagnetic order
respectively. Such order was quantified by the spin-spin correlation χ defined in Eq. (4.12). Color
denotes spin magnitude. White dashed lines: domain boundaries. Figures modified and reproduced
from [25]

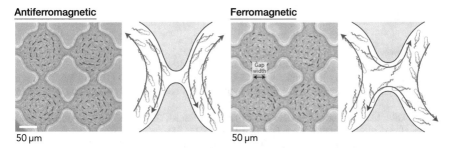

Fig. 4.7 Explanations for the emergence of ferromagnetic and antiferromagnetic order depending on the gap widths. Bacteria swimming along the wall forms edge currents, which drive internal vortical flow in cavities. For small gap widths, bacteria cannot easily cross the channel, but for large gap widths, they can swim across the cavities. Whether such edge currents dominate hydrodynamic continuity or not modifies the emergent order. Figures modified and reproduced from [25]

currents then drive internal vortices. In the case of large gap widths, bacteria can easily swim through the gap, and then the edge currents drive bulk flow in adjacent cavities in the same direction. On the other hand, for small gaps, it is quite difficult for bacteria to cross the channel to reach the adjacent cavities. Here, due to hydrodynamic continuity, the opposite rotations in adjacent cavities are more stable.

4.4.3 Summary and Interpretations

Through the experiments above, it has turned out that bacterial turbulence can sometimes self-organize in some sort of ordered collective states by imposing boundary conditions. It is interesting to know that we can still observe order in chaotic bacterial turbulence.

However, these experiments are done in confined geometry. What was explored in the second experiment [25] was the dynamics of connected bacterial vortices. As in the first experiment [23], bacteria inside each cavity were originally forced to form vortices due to boundaries. Therefore, it remained unexplored what happens when *unconstrained* bulk bacterial turbulence meets some structured obstacles. Especially, it was not clear how the size or the periodicity of structures affects the dynamics of bacterial turbulence.

Furthermore, such microfluidic experiments with dense suspensions of aerobic *Bacillus subtilis* cannot be run for long due to the shortage of oxygen. Bacterial turbulence rapidly decelerates and it is difficult to perform long measurements in steady states. As a matter of fact, the analysis in [25] was done with only 10-second movies. They could not observe the switching dynamics of spins or persistence of vortices.

Being aware of these problems, we performed our experiments on bacterial turbulence that will be presented in Chap. 5.

References

1. Dombrowski C, Cisneros L, Chatkaew S, Glodstein RE, Kessler JO (2004) Self-concentration and large-scale coherence in bacterial dynamics. Phys Rev Lett 93(9):098103
2. Cisneros LH, Cortez R, Dombrowski C, Goldstein RE, Kessler JO (2007) Fluid dynamics of self-propelled microorganisms, from individuals to concentrated populations. Exp Fluids 43(5):737–753
3. Dunkel J, Heidenreich S, Drescher K, Wensink HH, Bär M, Goldstein RE (2013) Fluid dynamics of bacterial turbulence. Phys Rev Lett 110(22):228102
4. Sokolov A, Aranson IS, Kessler JO, Goldstein R (2007) Concentration dependence of the collective dynamics of swimming bacteria. Phys Rev Lett 98(15):158102
5. Wensink HH, Dunkel J, Heidenreich S, Drescher K, Goldstein RE, Löwen H, Yeomans JM (2012) Meso-scale turbulence in living fluids. Proc Natl Acad Sci U S A 109(36):14308–14313
6. Creppy A, Prand Olivier, Druart X, Kohnke PL, Plouraboué F (2015) Turbulence of swarming sperm. Phys Rev E 92(3):032722
7. Nishiguchi D, Sano M (2015) Mesoscopic turbulence and local order in Janus particles self-propelling under an ac electric field. Phys Rev E 92(5):052309
8. Helbing D, Johansson A, Al-Abideen H (2007) Dynamics of crowd disasters: an empirical study. Phys Rev E 75(4):046109
9. Purcell EM (1977) Life at low Reynolds number. Am J Phys 45:3
10. Sokolov A, Aranson IS (2012) Physical properties of collective motion in suspensions of bacteria. Phys Rev Lett 109(24):248109
11. Nishiguchi D, Nagai KH , Chaté H, Sano M (2017) Long-range nematic order and anomalous fluctuations in suspensions of swimming filamentous bacteria. Phys Rev E 95(2):020601(R)
12. Drescher K, Dunkel J, Cisneros LH, Ganguly S, Goldstein RE (2011) Fluid dynamics and noise in bacterial cell–cell and cell–surface scattering. Proc Natl Acad Sci USA 108:10940–10945
13. Landau LD, Lifshitz EM (1987) Fluid mechanics, 2nd edn. Elsevier
14. Wensink HH, Löwen H (2012) Emergent states in dense systems of active rods: from swarming to turbulence. J Phys : Condens Matter 24(46):464130
15. Grossmann R, Romanczuk P, Bär M, Schimansky-Geier L (2014) Vortex arrays and active turbulence of self-propelled particles. Phys Rev Lett 113(25):258104
16. Bratanov V, Jenko F, Frey E (2015) New class of turbulence in active fluids. Proc Natl Acad Sci USA 112(49):15048–15053
17. Nishiguchi D, Iwasawa J, Jiang H-R, Sano M (2018) Flagellar dynamics of chains of active Janus particles fueled by an AC electric field. New J Phys 20:015002
18. Marenduzzo D, Orlandini E, Cates ME, Yeomans JM (2007) Steady-state hydrodynamic instabilities of active liquid crystals: hybrid lattice Boltzmann simulations. Phys Rev E 76(3):031921
19. Dunkel J, Heidenreich S, Bär M, Goldstein RE (2013) Minimal continuum theories of structure formation in dense active fluids. New J Phys 15:045016
20. Lauga E, DiLuzio WR, Whitesides GM, Stone HA (2006) Swimming in circles: motion of bacteria near solid boundaries. Biophys J 90(2):400–412
21. Lemelle L, Palierne JF, Chatre E, Place C (2010) Counterclockwise circular motion of bacteria swimming at the air-liquid interface. J Bacteriol 192(23):6307–6308
22. Figueroa-Morales N, Miño GL, Rivera A, Caballero R, Clément E, Altshuler E, Lindner A (2015) Living on the edge: transfer and traffic of E. coli in a confined flow. Soft Matter 11:6284–6293
23. Wioland H, Woodhouse FG, Dunkel J, Kessler JO, Goldstein RE (2013) Confinement stabilizes a bacterial suspension into a spiral vortex. Phys Rev Lett 110(26):268102
24. Lushi E, Wioland H, Goldstein RE (2014) Fluid flows created by swimming bacteria drive self-organization in confined suspensions. Proc Natl Acad Sci USA 111(27):9733–9738
25. Wioland H, Woodhouse FG, Dunkel J, Goldstein RE (2016) Ferromagnetic and antiferromagnetic order in bacterial vortex lattices. Nat Phys 12:341–345

Chapter 5
Encounter of Bacterial Turbulence with Periodic Structures

Abstract In this chapter, we report on self-organization of concentrated suspensions of motile bacteria *Bacillus subtilis* constrained by two-dimensional (2D) periodic arrays of microscopic vertical pillars. We show that bacteria self-organize into a lattice of hydrodynamically bound vortices with a long-range antiferromagnetic order controlled by the pillars' spacing. The patterns attain their highest stability and nearly perfect order for the pillar spacing comparable with an intrinsic vortex size of unconstrained bacterial turbulence. Even a small number of periodic obstacles, with the volume fraction as small as 4%, can trigger the emergence of stable long-range antiferromagnetic vortex lattices from such chaotically fluctuating bacterial turbulence. This strategy can be used to control a wide class of active systems. This chapter is based on our publication [Nishiguchi et al. Nature Communications, 9, 4486 (2018)] but with detailed explanations.

Keywords Bacterial turbulence · Direct laser lithography · Microscopic pillars · Emergent order · Antiferromagnetic order

5.1 Introduction

As we have seen in Chap. 4, properties of active turbulence formed by dense suspensions of bacteria—bacterial turbulence—have been investigated. Especially, its bulk properties have been clarified in detail experimentally [1–7], numerically [6, 8, 9], and theoretically [6, 7, 10]. Recently, the studies on active turbulence are moving forward to the next stage: How can we extract 'order' from spatio-temporally chaotic active turbulence? Along this line, it has turned out that bacterial turbulence confined in microfluidic chambers sometimes shows directed motion such as spontaneous spiral vortex formation [11, 12], ferromagnetic and antiferromagnetic vortex lattice formation [13], directed collective motion under channel confinement [14], etc.

However, it remains elusive and numerically/theoretically inaccessible how bulk unconstrained bacterial turbulence behaves when it encounters some obstacles and

© Springer Nature Singapore Pte Ltd. 2020 97
D. Nishiguchi, *Order and Fluctuations in Collective Dynamics*
of Swimming Bacteria, Springer Theses, https://doi.org/10.1007/978-981-32-9998-6_5

structures. Such studies are required to fill the gap between existing studies on bulk bacterial turbulence and confined bacterial turbulence.

Here we present our experimental study that treats the interplay between bacterial turbulence and periodic obstacles [15]. We rectify a turbulent dynamics in suspensions of swimming bacteria *Bacillus subtilis* by imposing periodical constraints on bacterial motion. Bacteria, swimming between periodically placed microscopic vertical pillars, self-organize in a stable lattice of vortices. We demonstrate the emergence of a strong antiferromagnetic order of bacterial vortices in a rectangular lattice of pillars. The emerged vortex lattices demonstrate only antiferromagnetic order, while a system of connected chambers exhibits the transition to a short-range ferromagnetic order with the increase of gap sizes [13]. We explain this apparent discrepancy by a different mechanism of interactions between the vortices in these two systems. Since the pillars are only 14-μm-wide, the bacteria are not able to swim along liquid-solid boundaries of the pillars and to self-organize in a stable circulating loop around each pillar. Correspondingly, the dynamics of this system cannot be described by dual interacting vortex lattices, between and around pillars, as in Ref. [13]. Instead, an array of tiny pillars creates a periodic set of stationary points with zero bacterial velocities. The emerged pattern arises from the continuity of bacterial flow between the pillars. Hydrodynamic interaction between vortices increases the stability of an emerged pattern. The highest stability of vortices in the antiferromagnetic lattice and the fastest vortices speed were observed in structures with the periods comparable with a correlation length of bacterial unconstrained motion. We constructed a coarse-grained model and explained the stabilization of the emergent antiferromagnetic vortex lattices. The obtained results highlight the existence of characteristic length scale in bacterial turbulence and its importance for the emergence of order out of chaos. This suggests a new less-invasive methodology to control and rectify the bacterial behavior at microscopic scales.

5.2 Experimental Procedure and Setup

We investigate the properties of suspensions of swimming bacteria *Bacillus subtilis* in the presence of periodic arrays of microscopic vertical pillars. The bacteria (strain 1085) were grown in a Terrific Broth (TB) medium and concentrated by centrifugation at the final concentration of $\sim 10^{10}$ cm^{-3}. A small drop of concentrated suspension is placed on a glass slide with the array of microscopic vertical pillars in such a way that air-liquid interface is a few microns below top surfaces of pillars. The drop on a slide was enclosed by a plastic spacer and a microscope coverslip with an air gap ≈ 0.5 mm. The enclosure minimized evaporation of water still providing oxygen to bacteria. After enclosure, the whole experimental cell is inverted so that the bacteria accumulate at the surface of the suspension due to gravity and aerotaxis. The dynamics of bacteria was captured by an Olympus IX71 inverted microscope and a high-resolution (5120×3840) HS20000C camera at $10\times$ magnification at 32.7 fps. We used only the red pixels of this RGB color camera for analysis because the images acquired by the red pixels (longer wavelength) had the highest spatial resolution due to the smaller amount of scattering and diffraction of transmitted light. By enlarging aperture stop of the microscope and decreasing the focus depth, we captured the turbulent dynamics of bacteria only at the surface. The experiments could run for minutes, but the speed of bacterial turbulence gradually decelerates. Therefore, we used the first 61 s of movies (2000 frames) for analysis, in which we have confirmed that the properties of unconstrained turbulence (the red region in Fig. 5.3) could be regarded as steady (see Appendix in Sect. 5.5.1).

The pillars were 3D-printed by direct laser lithography [16] on Photonic Professional GT system from Nanoscribe GmbH with the spatial resolution of 0.5 μm. Being 150-μm-tall and 20-μm-wide in diagonal lines, the pillars are arranged in 9 square lattices of the period a ranging from 50 to 130 μm with 10-μm increment (Fig. 5.1). The central part of the experimental cell is free of pillars. This area was used to measure parameters of unconstrained bacterial motion as a reference. To avoid a weird meniscus at the periphery of the suspension that affects the flatness of the observed surface and the behavior of outer regions, the region of printed pillars was surrounded by a 3D-printed wall with large tunnels through which bacteria can freely swim. In our experiments, we were able to track the dynamics of bacterial suspension simultaneously in all lattices with different pillars' spacing a. That significantly reduces the noise of collected data associated with variations of bacterial swimming speed, length, or fitness and age in different bacterial colonies, and makes it possible to reliably investigate how the period of structures affects the dynamics of bacterial turbulence.

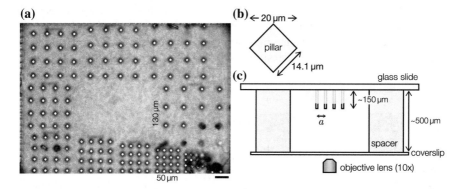

Fig. 5.1 a Snapshot of our experiment. Sets of 4×4 pillar lattices are constructed on a glass substrate, and bacterial turbulence was introduced so that bacteria can swim between the pillars. Pillar sets are organized in 9 arrays with different lattice constants a increasing from 50 to 130 μm (clockwise). The smallest lattice with $a = 40$ μm at the right-bottom of the field of view is excluded due to damaged pillars. The central region is left for bulk bacterial turbulence. Contrast is adjusted. Scale bar: 50 μm. **b** Schematic of a pillar. It has a square shape with the diagonal length 20 μm. **c** Schematic of our experimental setup. For clarity, only one set of pillars is depicted. The lattice constant a ranges from 50 to 130 μm. See also the Supplementary Movie 1 in Ref. [15]

The velocity field $v(r, t)$ of bacteria was obtained by using custom particle image velocimetry (PIV) MATLAB scripts (Fig. 5.2b). The PIV subwindows were 32 × 32 pixels (20 μm × 20 μm) and separated by every 8 pixels (75% overlap). The spatial resolution was smaller than any characteristic scales of observed collective motion. Estimation of bacteria velocity in the proximity of pillars is a technically challenging problem due to several effects. A meniscus creates an optical distortion in the vicinity of each pillars complicating bacteria tracking. In addition, the imposed geometrical confinements on bacterial motion near the pillars lead to a significant vertical motion and reduce the accuracy of spatial tracking. To avoid this problem, we excluded areas around each pillar from our analysis and measured all bacterial swimming parameters only in square regions of interest (ROI) between pillars, see Fig. 5.2b. We also excluded ROIs adjacent to distorted pillars, and ROIs shown in Fig. 5.3 are used for the following analysis.

5.3 Results

5.3.1 Antiferromagnetic Vortex Lattice

We observed the emergence of stable lattices of bacterial vortices. As shown in Fig. 5.2a, the time-averaged magnitude of the vorticity field $\langle \text{rot } v(r, t) \rangle_t$ demonstrates emergence of antiferromagnetic order between $a = 60$ μm and $a = 90$ μm.

In contrast with the recent work by Wioland et al. [13], the volume/area fraction of obstacles is much smaller, imposing effectively much less constraints on bacterial motion. Wioland et al. confined dense bacterial turbulence in the lattice of connected circular cavities. Each cavity imposes strong constraints on bacterial suspensions and triggers rotational motion inside itself [13]. In contrast, in spite of the tiny

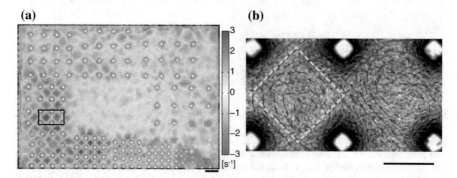

Fig. 5.2 **a** Color plot of the magnitude of the vorticity field $\langle \text{rot } v(r, t) \rangle_t$. **b** Close up of the rectangular area shown in (**a**). Arrows indicate instantaneous velocities. Yellow dashed line depicts a single ROI area. Scale bars: 50 μm. See also the Supplementary Movie 2 in Ref. [15]

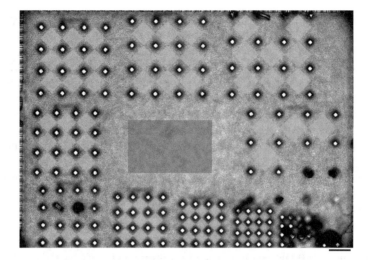

Fig. 5.3 The square ROIs used for analysis. Green regions for pillar lattices, and the red region for unconstrained bulk bacterial turbulence. Scale bar: 50 μm

size of pillars in our setup, we observed the emergence of stable lattices of bacterial vortices (Fig. 5.2a, b). Effectively large gaps between pillars increased hydrodynamic bindings between vortices.

However, surprisingly, the emerged lattice demonstrate only antiferromagnetic order and we could not observe ferromagnetic order in any case (Fig. 5.2a, b), in a seeming contradiction with previous results [13]. If we naively compare our results with those in [13], our setup corresponds to the large gap width case in [13]. In that case, Wioland et al. obtained ferromagnetic order, which is different from antiferromagnetic order in our case. This paradox can be explained by a different mechanism of interaction between vortices inside cavities [13] and between pillars (our case). Since the pillars are only 20 μm wide in diagonal lengths and have square shapes, the bacteria are not able to self-organize in a stable circulating loop around each pillar. Correspondingly, the dynamics of this system cannot be described by dual interacting lattices, between and around pillars [13]. Instead, an array of tiny pillars creates a periodic set of stationary points with zero bacterial velocities, while the velocity field between pillars changes continuously.

5.3.2 Vorticity and Enstrophy

To capture the properties of the emerged dynamical patterns, we measured the absolute values of mean vorticity $\langle |\langle \text{rot } v(r, t) \rangle_t | \rangle_{r \in \text{ROI}_a}$ and enstrophy $\langle \langle [\text{rot } v(r, t)]^2 \rangle_t \rangle_{r \in \text{ROI}_a}$ of the bacterial velocity field in ROIs. Here, ROI_a denotes a set of ROIs with the lattice constant a. For small lattice constants $a < 60$ μm, the bacteria are not able to develop turbulent motion, and their collective swimming is suppressed by densely placed pillars. As the lattice constant a increases, both the mean vorticity and the mean enstrophy increase, reaching maximum at $a = 60$–90 μm (Fig. 5.4a). For lattice periods of $a = 60 - 90$ μm, we observed an emergence of stable anti-ferromagnetic lattices of vortices. These lattice periods are comparable with a doubled correlation length or the scale of flows and vortices observed in unconstrained (unbounded) bacterial suspension. The stability of vortices is characterized by large magnitudes of vorticity and small temporal fluctuations (Fig. 5.4b). For large periods $a > 100$ μm, the bacterial suspension is quasi-turbulent, while the influence of sparsely placed pillars is reduced. The mean vorticity is reduced due to large temporal fluctuations, while the mean enstrophy remains almost a constant. Remarkably, in a previous study, bacteria swimming in a microfluidic confined channel exhibit a sharp transition from stable flow to a turbulent state with the increase of the channel width over ≈ 70 μm [14]. This as well as our results highlights the importance of this characteristic scale for both fully enclosed and slightly confined systems, and suggests a new methodology to efficiently control and rectify bacterial behavior at microscopic scales.

We characterized the temporal stability of vortices by calculating temporal fluctuations of velocity $\boldsymbol{v}_f$ and fluctuations of the tangential component of velocity $\boldsymbol{v}_t$ in a vortex for each lattice constant a,

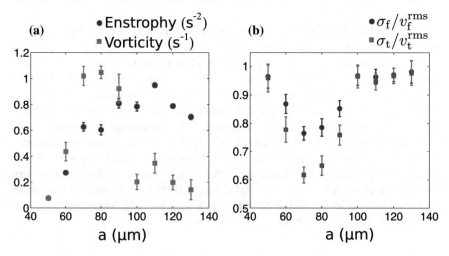

Fig. 5.4 **a** Mean enstrophy $\langle\langle[\mathrm{rot}\, \boldsymbol{v}(\boldsymbol{r}, t)]^2\rangle_t\rangle_{\boldsymbol{r}\in\mathrm{ROI}_a}$ (blue circles) and absolute values of mean vorticity $\langle|\langle\mathrm{rot}\, \boldsymbol{v}(\boldsymbol{r}, t)\rangle_t|\rangle_{\boldsymbol{r}\in\mathrm{ROI}_a}$ (red square) as a function of lattice constant a. **b** Dependence of temporal fluctuations of full velocity $\sigma_f/v_f^{\mathrm{rms}}$ (blue circles) and tangent velocity $\sigma_t/v_t^{\mathrm{rms}}$ (red squares) on the lattice constant a. Both fluctuations have minima at 70 μm, at which the vortices exhibit stable antiferromagnetic order. Error bars are estimated as the standard deviations among ROIs with the same a

$$\sigma_{f,t}(a) = \left\langle \sqrt{\left\langle [\boldsymbol{v}_{f,t}(\boldsymbol{r}, t) - \langle \boldsymbol{v}_{f,t}(\boldsymbol{r}, t)\rangle_t]^2 \right\rangle_t} \right\rangle_{\boldsymbol{r} \in \mathrm{ROI}_a} \tag{5.1}$$

The tangential component of velocity is measured relative to the center of each ROI. Since the average magnitude of velocity is different for each a, we normalized σ_f and σ_t by the root mean square (rms) velocities v_f^{rms} and v_t^{rms} respectively for each a,

$$v_{f,t}^{\mathrm{rms}}(a) = \left\langle \sqrt{\langle [\boldsymbol{v}_{f,t}(\boldsymbol{r}, t)]^2 \rangle_t} \right\rangle_{\boldsymbol{r} \in \mathrm{ROI}_a}. \tag{5.2}$$

Normalized fluctuations are shown in Fig. 5.4b as functions of the pillar lattice constant a. Both $\sigma_f/v_f^{\mathrm{rms}}$ and $\sigma_t/v_t^{\mathrm{rms}}$ exhibit minima at $a = 70$ μm. At this lattice constant value, the vortices are hydrodynamically stabilized and the bacterial suspension self-organizes in the most stable coherent vortex lattice. As we have already mentioned above, the vorticity field $\langle \mathrm{rot}\, \boldsymbol{v}(\boldsymbol{r}, t)\rangle_t$ calculated from the PIV clearly demonstrates the antiferromagnetic order inside the pillar arrays for the lattice spacing a between 60 and 90 μm (Fig. 5.2a). For larger periods, $a > 100$ μm, the bacterial suspension is quasi-turbulent due to the reduced influence of the pillars. The mean vorticity is reduced as well due to large temporal fluctuations, while the mean enstrophy remains almost constant.

Error bars represent standard errors estimated from standard deviations among the ROIs in the same lattice. For quantities $F(\boldsymbol{r})$ such as vorticity $|\langle \mathrm{rot}\, \boldsymbol{v}(\boldsymbol{r}, t)\rangle_t|$ before taking average $\langle\ \rangle_{\boldsymbol{r} \in \mathrm{ROI}_a}$ and fluctuations $\sqrt{\langle [\boldsymbol{v}_{f,t}(\boldsymbol{r}, t) - \langle \boldsymbol{v}_{f,t}(\boldsymbol{r}, t)\rangle_t]^2 \rangle_t}$, error bars are calculated as,

$$\sqrt{\frac{\sum_i \left[\langle F(\boldsymbol{r})\rangle_{\boldsymbol{r} \in \mathrm{ROI}_a^i} - \langle F(\boldsymbol{r})\rangle_{\boldsymbol{r} \in \mathrm{ROI}_a} \right]^2}{N(a) - 1}}, \tag{5.3}$$

where ROI_a^i denotes the i-th ROI in the lattice with the period a, and $N(a)$ is the number of the analyzed ROIs in the lattice with the period a.

5.3.3 Antiferromagnetic Order Parameter

The antiferromagnetic order of vortices can be quantified with a spin-spin correlation, similar to what was introduced in the early study [13]. We introduce a spin variable for each ROI in the lattice at time t,

$$S_{i,a}(t) := \frac{\hat{\boldsymbol{z}} \cdot \left[\sum_{\boldsymbol{r} \in \mathrm{ROI}_a^i} (\boldsymbol{r} - \boldsymbol{r}_i) \times \boldsymbol{v}(\boldsymbol{r}, t) \right]}{\sum_{\boldsymbol{r} \in \mathrm{ROI}_a^i} |\boldsymbol{r} - \boldsymbol{r}_i|}, \tag{5.4}$$

where r_i is the geometrical center of ROI_a^i, and $\hat{z}$ is the unit vector in the vertical direction. The magnitude of the spin represents the relative strength of a vortex and the sign of the spin reflects the predominant direction of rotation, positive for counter-clockwise and negative for clockwise. For the ferromagnetic order, signs of neighbor spins are the same, while for antiferromagnetic order the signs are alternating along both main lattice axes. To introduce the order parameter, we calculated the adjacent spin correlation $\chi_a(t)$ for each lattice constant a,

$$\chi_a(t) := \frac{\sum_{i \sim j} S_{i,a}(t) S_{j,a}(t)}{\sum_{i \sim j} |S_{i,a}(t) S_{j,a}(t)|}, \qquad (5.5)$$

where the sum $\sum_{i \sim j}$ runs over all adjacent pairs in a single lattice structure. We applied Savizky-Golay filter with degree 3 and a duration of 0.3 s (11 frames) to eliminate high-frequency noise associated with PIV and data processing errors just for visualization of the time series as presented in Fig. 5.5a.

For a small lattice period $a = 50\,\mu\text{m}$, the order parameter fluctuates near -0.25 and occasionally drops to ≈ -0.75. Densely placed pillars decrease average bacterial swimming velocity and prevent self-organization into a stable lattice. The observed antiferromagnetic order for $a = 60$–$90\,\mu\text{m}$ is characterized by strong anti-correlation between the adjacent vortices, $\chi_a \approx -1$. For $a > 100\,\mu\text{m}$, at the quasi-turbulent regime, the average order parameter $\langle \chi_a \rangle_t$ is relatively small, but temporal fluctuations of $\chi_a(t)$ are large as shown in Fig. 5.5a. For a short period of time, the swimming bacteria self-organize into a large-scale coherent structure with a characteristic scale of the order of the lattice period. However, since the characteristic scale of such motion is larger than a preferable vortex size, large vortices quickly break down into smaller vortices with a more favorable scale of ~ 60–$80\,\mu\text{m}$, resulting in large fluctuations of spins and chaotic hydrodynamic interaction between adjacent ROIs. Importantly, we did not observe any tendency to self-organize in a coherent structure for $a = 130\,\mu\text{m}$ which is roughly the doubled period of the most stable lattice constant $a = 70\,\mu\text{m}$ and intrinsic length scale of vortices. That emphasizes a high sensitivity of bacterial vortices pattern to defects in pillar lattices.

5.3.4 Persistence of Vortices: Life Times

We investigated temporal properties of the observed patterns. Since the period of the antiferromagnetic lattice of bacterial vortices is equal to the doubled period of the pillar lattice, the stable antiferromagnetic lattice has two possible spatial configurations. One configuration transforms to the other by shifting along the main axes by a. Although such a transition requires simultaneous sign flipping of all spins in a single lattice and is hardly observable in experiments, occasionally, a single bacterial vortex may switch the direction of rotation or flip the spin (Fig. 5.6b). An important question here is how the vortex size and the spin-spin interaction in a lattice of different period

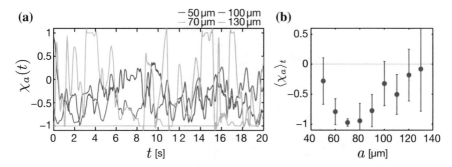

Fig. 5.5 a Temporal dynamics of spin correlations $\chi_a(t)$ for different lattice constants a represented by different colors. **b** Time-averaged values of adjacent spin correlations $\langle\chi_a(t)\rangle_t$ versus the lattice constant a. Strong antiferromagnetic order ($\langle\chi_a(t)\rangle_t \simeq -1$) is observed around $a \simeq 70\,\mu$m. Error bars: standard deviations of time series of $\chi_a(t)$

affect the stability of a single vortex. To answer this question, we measured the mean persistence time or life time of the vortices in arrays of different periods.

Each vortex in a pillar array is bound with 4 adjacent vortices, while a vortex on the edge interacts with a turbulent bacterial bath and therefore is less stable. To minimize the influence of such fluctuations, we fabricated larger arrays of pillars (9×9 instead of 4×4) for periods $a = 50$–$90\,\mu$m, see Fig. 5.7a, b and Supplementary Movies 3, 4, 5 in Ref. [15]. To eliminate the boundary effects for obtaining the probability $P_a(t)$ accurately, we excluded the outer two layers of the ROIs from the analysis. In this experiment, we captured the movie at 32 fps.

Existence of stable lattices of interacting spins complicates the spin switching dynamics and requires careful analysis. Because the vortex lattices for $a = 60$–$90\,\mu$m is stable, there are favorable and unfavorable directions of rotation for each ROI in these lattices. We do observe that temporal fluctuations lead to rotations with unfavorable directions, but such rotations are short-lived compared with ones with favorable directions. Therefore, there exist two distinct life times τ_a^{long} and τ_a^{short} corresponding to favorable and unfavorable directions respectively. In other words, due to the hydrodynamic interaction between vortices, the short time τ_a^{short} corresponds to switching from local ferromagnetic (unstable) order to antiferromagnetic (stable) order (Fig. 5.6b). In a stable antiferromagnetic configuration, a vortex remains its orientation for a much longer period of time τ_a^{long}. By analyzing the spins dynamics, we obtained the probability $P_a(t)$ of a spin to remain oriented in the same direction for a time t. This probability drops quickly for $a = 50\,\mu$m and $a = 90\,\mu$m (Fig. 5.7c) due to chaotic behavior of the bacterial flow. The exponential decay of P_a is a manifestation of Poisson random process: The probability of switching remains the same for any given period of time. The spin persistence time τ_a was estimated by fitting the experimental data with $P_a(t) \propto \exp(-t/\tau_a)$. The fitting range was [0 s, 1 s] for unfavorable directions and [0 s, 15 s] for favorable directions. In the stable antiferromagnetic configuration for $a = 70\,\mu$m, each spin retains its favorable locally antiferromagnetic orientation for $\tau_a^{\mathrm{long}} \approx 40$ s, while very rarely flipping to unfavor-

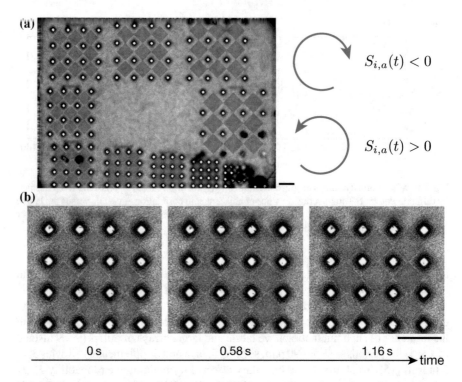

Fig. 5.6 Instantaneous signs of spins are overlaid on experimental snapshots. Clockwise rotations ($S_{i,a}(t) < 0$) and counterclockwise rotations ($S_{i,a}(t) > 0$) are represented by red and blue respectively. **a** Typical snapshot of the whole field of view. **b** Emergence of unstable short-lived ferromagnetic configuration at $a = 70$ μm. Scale bars: 100 μm

able locally ferromagnetic orientation for $\tau_a^{\text{short}} \approx 0.2$ s (Fig. 5.6b). This time τ_a^{short} is significantly shorter than a typical time scale of bacterial dynamics ~ 1 s. The magnitude of the spin fluctuates around zero for quasi-turbulent lattices, $a \geq 100$ μm.

As a result, the life time τ_a as a function of the period a clearly exhibits a sharp peak (Fig. 5.7d), which demonstrates that bacterial turbulence responds quite sensitively to the periodic structures at around 70 μm.

5.3.5 Correlation Function of Velocity Field

Interaction of swimming bacteria inside the same ROI can be quantified by the spatial correlation function $C_a(r)$ and the correlation length L_a of the velocity field $\boldsymbol{v}(\boldsymbol{r}, t)$. The correlation function was calculated in each i-th ROI for each lattice constant a and then averaged over both i and time,

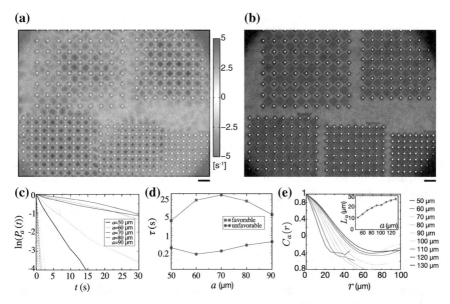

Fig. 5.7 Stability of antiferromagnetic order and velocity correlation functions. **a, b** Experimental snapshot of large arrays (9×9) of pillars for $a = 50$–90 μm with a step of 10 μm overlaid with (**a**) average vorticity of the bacterial velocity field and **b** instantaneous signs of the spins. Some pillars are out of the field of view. A perimeter of vortex lattices is excluded from further image processing, leaving the internal region of the size 4×4 vortices for analysis. Scale bar is 100 μm. **c** Semi-log plots of the vortex persistence probability $P_a(t)$ for different lattice constants a. Solid and dashed lines correspond to probabilities of a vortex to remain in antiferromagnetic and ferromagnetic orientation correspondingly. **d** Semi-log plots of persistence times of a vortex as functions of the lattice constant. For $a = 60$–80 μm, stable antiferromagnetic order significantly increases the persistence time τ_a^{long} for the preferred antiferromagnetic vortex state (red squares) and decrease the time τ_a^{short} for more unfavorable local ferromagnetic orientation (blue circles). **e** Correlation functions of the velocity field $\boldsymbol{v}(\boldsymbol{r}, t)$ inside the lattice structures with different lattice constants. Inset: Correlation lengths as a function of the lattice constant a. The correlation length in unconstrained suspension is $L_\infty \simeq 45$ μm

$$C_a(r) := \frac{\langle\langle\langle \boldsymbol{v}(\boldsymbol{r}', t) \cdot \boldsymbol{v}(\boldsymbol{r}' + \boldsymbol{r}, t)\rangle_{\boldsymbol{r}', \boldsymbol{r}' + \boldsymbol{r} \in \text{ROI}_a^i}\rangle_i\rangle_t}{\langle\langle\langle |\boldsymbol{v}(\boldsymbol{r}', t)|^2\rangle_{\boldsymbol{r}' \in \text{ROI}_a^i}\rangle_i\rangle_t}. \tag{5.6}$$

As expected, the correlation function $C_a(r)$ decays with r (Fig. 5.7e). The presence of a vortical type of motion inside each ROI is manifested by negative values of C_a at large r, which is especially noticeable for $a = 70$–90 μm. The increase of $C_a(r)$ with r for $r > 70$ μm is observed in the pillar lattices with large periods $a > 100$ μm. It indicates that diameters of observed vortices are smaller than the periods of pillar lattices and that they are ≈ 70 μm. Therefore, when $a > 70$ μm, more than one vortices can coexist in a single ROI, which results in frustrated configurations of vortices and leads to destabilization of the antiferromagnetic vortex order.

The correlation length L_a is defined as the distance at which $C_a(r)$ becomes smaller than $1/e$. L_a increases with a as expected and approaches the correlation length of unconstrained suspension $L_\infty \simeq 45$ μm (Fig. 5.7e).

5.3.6   Hexagonal Lattice

While the main focus of our work was on the dynamics of swimming bacteria in the square lattices of pillars, we also performed additional experiments for hexagonal lattices (Fig. 5.8). As we can infer from the results on square lattice experiments, hydrodynamic continuity cannot be fulfilled without imposing any frustration between spins in hexagonal lattices. In accordance with the lattice geometry, we printed hexagonal pillars for this experiment instead of square pillars used in the square lattice experiments.

Because we have obtained the strong antiferromagnetic order in the square lattices with the lattice constant $a \simeq 70$ μm, we investigated hexagonal lattices whose diagonal distance is comparable to 70 μm. In these hexagonal lattices, we defined the lattice constant a as the distance between two nearest pillars, and we chose appropriate lattice constants a so that the diameter of inscribed circles of hexagons $\sqrt{3}a \simeq 70$ μm, or $a \simeq 40.4$ μm. Therefore, we tested $a = 40$ μm and $a = 45$ μm. In this experiment, we captured a movie at 52.7 fps and analyzed 4112 frames (78 s).

In spite of such choice of length scale that is favorable to vortex formation, we could observe neither ferromagnetic nor antiferromagnetic order in hexagonal lattices (Fig. 5.9). We analyzed the adjacent spin correlation $\chi_a(t)$ defined in Eq. (5.5) by defining the ROIs as shown in Fig. 5.10a. The spin correlation $\chi_a(t)$ fluctuates a

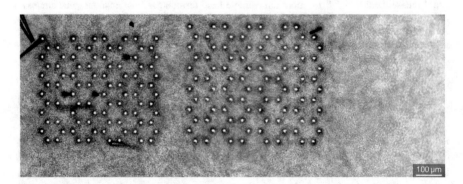

Fig. 5.8 Snapshot of the experiments on hexagonal lattices. The distances between the two nearest pillars, or the lattice constant a, are $a = 40$ μm for the left lattice and $a = 45$ μm for the right lattice

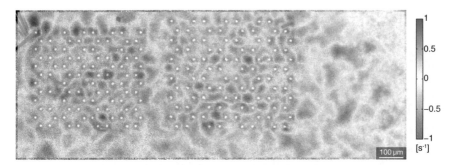

Fig. 5.9 Color plot of the time-averaged vorticity field overlaid on the snapshot of Fig. 5.8. Neither ferromagnetic nor antiferromagnetic order is observed as expected

lot for both $a = 40$ μm and $a = 45$ μm, but they almost always stay around 0 and $|\chi_a(t)| < 0.5$, which means that there is no stable emergent order.[1]

This again seemingly contradicts with the results in [13], in which they observed strong ferromagnetic order of bacterial turbulence confined in a hexagonal lattice of a microfluidic device, and highlight the importance of slight difference in boundary conditions for understanding macroscopic behavior, especially the emergent order, of bacterial turbulence. These results demonstrate again that the vortex lattice formation is triggered in accordance with hydrodynamic continuity conditions, and the underlying mechanism of vortex lattice formation is distinctively different from the previous study [13].

These results demonstrate again that the vortex lattice formation is triggered in accordance with hydrodynamic continuity conditions, and the underlying mechanism of vortex lattice formation is distinctively different from the previous study [13].

5.3.7 Self-organization of Swimming Bacteria Around Chiral Pillars

So far the studies where limited to non-chiral pillar arrays. Chirality is an important factor controlling the organization of collective bacterial motion. However, fabrication of chiral obstacles with controlled shapes used to be prohibitively difficult. Thanks to recent progress in two-photon photolithography, we 3D-printed arrays of hollow chiral towers, 30 μm in diameter and 100 μm in height. The tip of each tower has holes oriented at 45° (positive towers) or 135° (negative towers) to the radius of the tower (Fig. 5.11a). Bacteria, swimming from and to the center of each tower through these holes, create a vortex (Fig. 5.11b). The spin state of the vortex is prescribed by the "chirality" of the tower. Depending on the pre-manufactured order of a chiral tower array (antiferromagnetic or ferromagnetic), the bacterial suspension

---

[1]$0 \ll \chi_a(t) \leq 1$ and $-1 \leq \chi_a(t) \ll 0$ correspond to ferromagnetic and antiferromagnetic order respectively.

(a)

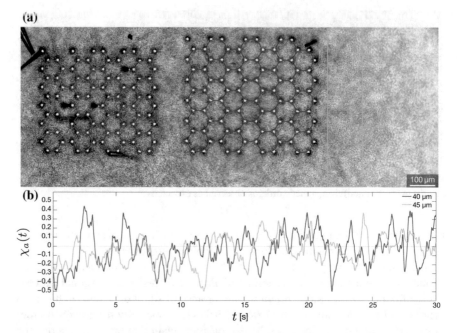

Fig. 5.10 **a** ROIs used for analysis. Red circular ROIs were used inside the lattices. Regions adjacent to distorted pillars were neglected for analysis. Green rectangular area represents the reference area for the unconstrained turbulence. **b** Temporal dynamics of spins $\chi_a(t)$ for different lattice constants a. Blue: $a = 40\,\mu$m. Orange: $a = 45\,\mu$m. The spin correlation $\chi_a(t)$ fluctuates around 0 by frequently changing its sign, and no clear order emerges

self-organizes into a stable antiferromagnetic state with a period of the tower lattice, a, or a "double-lattice" state according to the definition of Ref. [13] (Fig. 5.11c). The second state is represented by two ferromagnetic lattices of opposite spins shifted by a half-period along both crystallographic axes. By combining positive, negative and neutral (not chiral) towers in different patterns, we can produce various types of vortex lattices and control their stability.

5.3.8 Lattice Size Scaling Experiment

Fluctuations suppress a long-living antiferromagnetic order in relatively small arrays of pillars. The stability of any finite size lattice clearly depends on a fraction of vortices located on its perimeter, exposed to a turbulent bacterial bath. In contrast to the microfluidic chamber experiments with completely confined bacterial turbulence [11, 13, 17], the coexistence of the destabilizing bulk turbulence and the stable vortex lattices in our experiments enable us to assess such effects in detail.

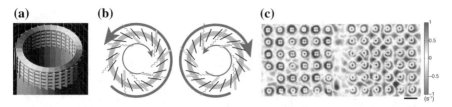

Fig. 5.11 Bacterial vortex lattices in the presence of chiral towers. **a** 3D model of chiral tower. The top part consists of five 3-μm thick hollow disks spaced at 3-μm vertical distance from each other. These disks are supported by 1-μm thin vertical walls oriented at 45° relative to the radius. **b** Chiral towers create internal and external hydrodynamical bacterial vortices due to specially oriented holes in their tips. **c** Time-averaged vorticity map of bacterial flow between antiferromagnetic (left) and ferromagnetic (right) chiral tower arrays. Scale bar: 100 μm

To assess how the stability of the antiferromagnetic vortex lattices depend on the size of the lattices or the proportion of the vortices on the lattice perimeters, we conducted experiments on pillar arrays with different sizes. We 3D-printed pillar arrays with the most stabilizing period $a = 70$ μm so that we can capture simultaneously the self-organized vortex lattices with their sizes $n \times n = 1 \times 1, 2 \times 2, \ldots, 7 \times 7$ as shown in Fig. 5.12. (Note that here n denotes the linear dimension of *vortex* lattices and that the sizes of the corresponding *pillar* arrays are $(n + 1) \times (n + 1) = 2 \times 2, 3 \times 3, \ldots, 8 \times 8$.) To reduce the interactions between vortex lattices in different pillar arrays, the pillar arrays were separated from each other by the distance sufficiently larger than the correlation length of the bulk bacterial turbulence ($L_\infty \simeq 45$ μm). Due to the limited field of view of the camera (1640 μm × 1230 μm), we captured two experimental movies at different positions with overlapping fields of view. We first captured the region shown in Fig. 5.12b (Movie I) and then the other region shown in Fig. 5.12a (Movie II). We used the same microscope and the same camera as the other previous experiments, and captured the movies at 30.0 fps for 133 s (4000 frames).

First, we calculated the average vorticity field and defined the favorable direction of rotation for each spin (Fig. 5.13a, b and Supplemental Movies 7–10 in Ref. [15]). Then instantaneous signs of spins in all the lattices are calculated (Fig. 5.13c, d) and used to evaluate the persistence times of the spins τ_k for their favorable directions in the same manner as described above. Here, k denotes the numbers of neighboring spins. Depending on the position in the lattice, k assumes a value of 2 on the vertices, 3 on the edges, and 4 in the bulk (Fig. 5.15b). We also printed an array of 4 pillars (just a single vortex) to attain the value of $k = 0$.

The dependence of the persistence time τ versus the number of neighbors k shows an exponential increase, consistent with our theoretical prediction based on the Kramers escape rate which will be presented in the following subsection. We calculated τ_k for the different lattice sizes (Fig. 5.14a), and then all the data for the same k were averaged (Fig. 5.15a). Error bars in Fig. 5.15a are standard errors calculated from the standard deviations among τ_k obtained from the different lattice sizes.

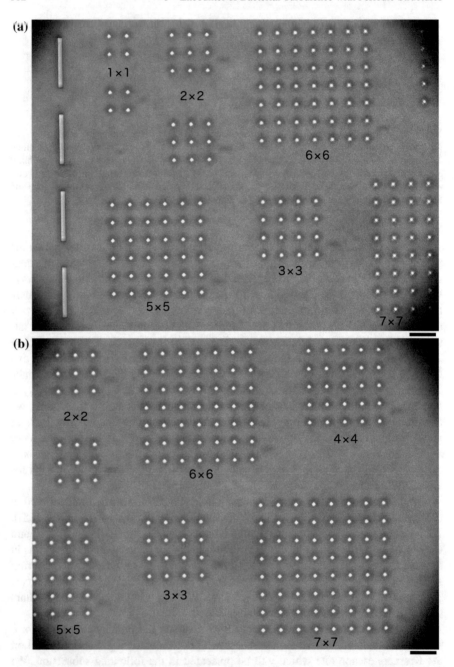

Fig. 5.12 Snapshots for lattice size scaling experiments. We captured two movies with overlapping fields of view successively: first Movie I for (**b**) and then Movie II for (**a**). As seen on the left side of (**a**), we 3D-printed walls surrounding the pillar arrays to avoid the effect of the meniscus. Scale bars: 100 μm

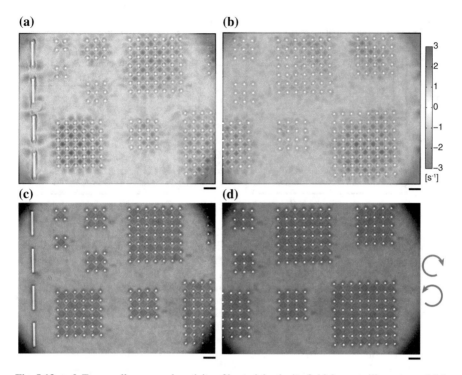

Fig. 5.13 a, b Temporally averaged vorticity of bacterial velocity field. Larger pillar arrays exhibit stronger antiferromagnetic vortex lattice formation. Color bar is common to (**a**) and (**b**). **c, d** Instantaneous signs of spins are overlaid on experimental snapshots in the same manner as in Fig. 5.6. Scale bars: 100 μm

Then we calculated the order parameter $\langle \chi_a \rangle_t$ for the whole lattice, including perimeters, for different array sizes from $n \times n = 2 \times 2$ to 8×8, while keeping the lattice constant $a = 70$ μm, see examples in Fig. 5.15d. The order parameters $\langle \chi_a \rangle_t$ obtained from Movie I and Movie II of the lattice size scaling experiment (Fig. 5.12) and the large lattice experiment (Fig. 5.7) are plotted together in Fig. 5.14b. The average values of these data are shown in Fig. 5.15c with the error bars calculated from the error propagation law using the errors in Fig. 5.14b. Since the peripheral vortices are more exposed to turbulent bath fluctuations, in small arrays of $n = 2$– 4, unstable vortices on the perimeter fluctuate frequently and destabilize the entire lattice. For larger arrays of $n = 5$–7, their influence is reduced and becomes negligible for $n \geq 8$. The dependence of the order parameter versus n is consistent with the $1/n$ law. The behavior can be inferred from the fact that the ratio of less coherent peripheral vortices to the bulk vortices scales as $1/n$.

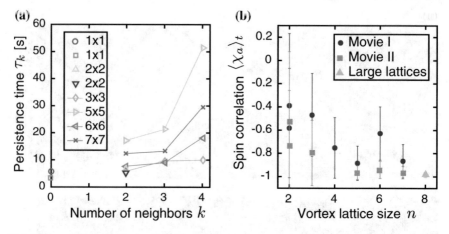

Fig. 5.14 a Persistence time τ_k of the favorable direction versus the number of neighbors k for the lattices with different sizes $n \times n$ calculated from Movie II. Average of these data is shown in Fig. 5.15a. **b** Order parameter (spin correlation) $\langle \chi_a \rangle_t$ versus the linear size n of the vortex lattice at the lattice constant $a = 70\ \mu m$. Data obtained from the lattice size scaling experiment (Movies I and II, blue circles and red squares respectively) and the large 9×9 pillar array (8×8 vortices) experiments (magenta triangles) are shown together. Average of these data is shown in Fig. 5.15c. Error bars: standard deviations of time series of $\chi_a(t)$

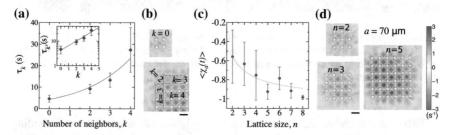

Fig. 5.15 Persistence time and order parameter dependencies. **a** Persistence time τ_k of a vortex with different numbers of neighbors k, shown in (**b**). Symbols depict experimental data, and the solid line is a fit to theoretical dependence $\tau_k \sim \tau_0 \exp(\alpha k)$, with $\tau_0 = 4.25$ s, $\alpha = 0.43$, see Eq. (5.12). Inset: τ_k versus k in semi-log scale. Error bars: standard errors of measurements. **b** Location of bacterial vortices with different numbers of neighbors k. Scale bar: $70\ \mu m$. Color bar is the same as that in (**d**). **c** Order parameter $\langle \chi_a(t) \rangle_t$ for a whole vortex lattice (including perimeter) of the size $n \times n$. Lattice constant $a = 70\ \mu m$. Symbols shows experimental data, solid line is a fit $\langle \chi_a(t) \rangle_t = -1 + 0.88/n$. Error bars are calculated from the standard deviations of time series $\chi_a(t)$. **d** Vorticity map of a bacterial flow for lattices of different sizes $n \times n$. Shown examples for $n = 2, 3, 5$. Lattice constant (distance between pillars) $a = 70\ \mu m$

5.3.9 Theoretical Description on Persistent Time

To estimate the persistence time as a function of the number of neighbors, we use a coarse-grained approach where individual spins are described by an angle variable

ϕ with the corresponding spin value $V = \cos(\phi)$. The angle ϕ is governed by the following equation,

$$\frac{d\phi}{dt} = -\gamma \sin(2\phi) + \xi(t), \qquad (5.7)$$

where γ is the relaxation rate, and $\xi(t)$ is a Gaussian white noise approximating interaction with the turbulent bacterial bath, $\langle \xi(t) \rangle = 0$, $\langle \xi(t)\xi(t') \rangle = 2D\delta(t - t')$ with D being the noise intensity. We note that the choice of the sine function in Eq. (5.7) is obviously not unique. Qualitatively similar results can be obtained for arbitrary symmetric bi-stable function.

In the absence of noise, a solution to Eq. (5.7) relaxes to either 0 or π. Correspondingly, the spin variable tends to $V = \pm 1$, representing clockwise/counterclockwise rotating vortices (compare to approach in Ref. [13]). With the noise, the system switches between these symmetric steady states. The persistence time τ_0 of an isolated spin can be estimated from the Kramers escape rate [18] (since the system needs to overcome the energy barrier γ to switch from the state 0 to π and vice versa),

$$\tau_0 \sim \exp(\gamma/D). \qquad (5.8)$$

Interacting spins ϕ_{ij} on a square lattice can be described similarly. Here the angle variable ϕ_{ij} represents the spin in the i-th row and j-th column in a square lattice. The governing equation can be written as,

$$\frac{d\phi_{ij}}{dt} = -\frac{dH}{d\phi_{ij}} + \xi_{ij}(t), \qquad (5.9)$$

where the free energy functional H is of the form

$$H = -\frac{1}{2}\gamma \sum_{ij} \cos(2\phi_{ij}) + \eta \sum_{ij} \sum_{lm} \cos(\phi_{ij} - \phi_{lm}). \qquad (5.10)$$

Here $\eta > 0$ is the interaction parameter, and the sums $\sum_{ij}$ and $\sum_{lm}$ are taken with respect to all nearest neighboring sites on a square lattice.

In the absence of noise, Eq. (5.9) favors a stable antiferromagnetic lattice corresponding to the minimum of the free energy H. With fluctuations, the spins flip, and the persistence time depends on the number of nearest neighbors k. One can find an upper bound for the energy barrier E versus k ($k = 0$ for an isolated spin, $k = 2$ for vertices, $k = 3$ for edges, and $k = 4$ inside the lattice as in Fig. 5.15b):

$$E \approx \gamma + 2\eta k. \qquad (5.11)$$

It gives the following estimate for the persistence time τ_k,

$$\tau_k \sim \tau_0 \exp(2\eta k/D) \qquad (5.12)$$

The exponential dependence for the persistence time τ_k given by Eq. (5.12) is in excellent agreement with the experiment (Fig. 5.15a).

To further explore the implication of our model, we calculated the persistence time τ_k for different lattice periods a by using the data from the large lattice ($n = 8$) experiments (Fig. 5.16a, b). Because our theoretical model assumes stable vortex lattice formation, it cannot be applied to relatively unstable lattices for $a = 50$ μm and $a = 90$ μm. In fact, the behavior of the persistence time τ_k for $a = 50$ μm and $a = 90$ μm clearly deviates from the exponential dependence (Fig. 5.16a, b). Therefore, we fitted the experimental data only for $a = 60, 70, 80$ μm where stable antiferromagnetic lattices were observed (Fig. 5.16c, d). The slopes of the exponential fitting on the semi-log scale correspond to $2\eta/D$ in Eq. (5.12), and the ratio of η and D has a very weak dependence on the lattice spacing a (Fig. 5.16e, f). Considering that the strength of noise D on the spin dynamics originates from the dynamics of the bacterial turbulence in the ROI and each spin is calculated by integrating the whole area of the corresponding ROI, we speculate that both η and D scale with the area of vortices: $\eta \propto a^2$ and $D \propto a^2$ resulting in $2\eta/D \approx$ const.

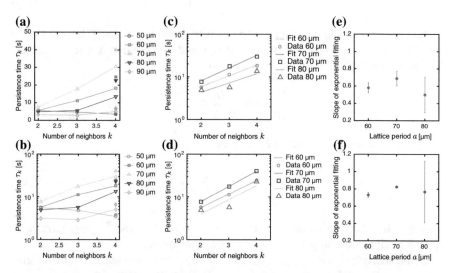

Fig. 5.16 a Persistence time τ_k versus the number of neighboring spins k for different lattice periods a. The data were obtained from the large lattice ($n = 8$) experiments shown in Fig. 5.7. At $k = 4$, two data points are plotted for each lattice period. The smaller ones with blank symbols were calculated from all the spins with $k = 4$ in the lattice, and the larger ones with filled symbols were calculated from the bulk spins by excluding the peripheral spins in the outer two layers of the lattices (same data as in Fig. 5.7d). **b** Semi-log plot of the same data as in (**a**). **c, d** Fitting results of the experimental data to the exponential behavior described in Eq. (5.12) by using (**c**) all the spin data and (**d**) the bulk spin data at $k = 4$. **e, f** Slopes of the exponential fittings in (**c**) and (**d**) respectively as functions of lattice period a. The slopes correspond to $2\eta/D$ in Eq. (5.12). Slopes do not change much, suggesting the similar dependence of both the interaction parameter η and the noise strength D on the lattice period a. Error bars: standard errors of the linear fitting on the semi-log scale

5.4 Conclusion

In conclusion, we have shown that dense suspensions of swimming bacteria self-organize into antiferromagnetic vortex lattices in the presence of periodic structures comparable with its intrinsic length scale. We have addressed what happens when bulk unconstrained bacterial turbulence encounters obstacles. Such an open system is a natural setup for bacteria rather than confined geometries if we think about the environment they live in. We have demonstrated that such chaotic turbulent phases can even self-organize into ordered motion just by imposing a small number of constraints taking only 3–5% of total suspension volume at the periodicity comparable with the correlation length.

This experimental observation of antiferromagnetic order is in contradiction with the naive comparison with the previous study on bacterial turbulence in confinement [13], suggesting a different mechanism for the bacterial vortices interactions. The difference can be explained by two possible reasons: (i) hydrodynamic continuity and (ii) absence of circulating swimming along pillars. In the microfluidic experiment [13], the size of the wall was larger than our pillars and bacteria can swim along the wall and exhibit circulating motion along the boundary. This circulating motion then drives vortices inside cavities, and this effect dominated hydrodynamic continuity at the large gap width case. However, in our pillar experiments, the size of the pillars is too small for bacteria to swim along the pillars. Therefore, the circulating swimming has little effect on the collective dynamics and the hydrodynamic continuity is dominant, leading to the antiferromagnetic order in the square lattices. This explanation is also consistent with the experimental results on hexagonal lattices.

Furthermore, our experiments could run for a much longer time than previous experiments in microfluidic confinement. Typical durations of such previous experiments were shorter than 10 s [11, 13], because bacterial turbulence rapidly decelerates due to the shortage of oxygen. On the other hand, in our setup, bacterial suspension is in contact with the air and our experiments could run for minutes. Although we restricted ourselves to use only the first 1-minute movies to avoid possible changes of bacterial dynamics, our longer observation enabled us to extract stability of vortex lattices in terms of persistence probability and life times, which had not been experimentally accessible with microfluidic devices. The observed patterns demonstrated significantly higher persistence and much larger magnitude of the order parameter in spite of small volume/area of pillars: Adjacent spin-spin correlation achieves the value close to -1 for $a = 70\,\mu$m. This high vortex lattice robustness prevents penetration of defects from the perimeter of the lattice to the bulk. For the lattices with the $a = 70\,\mu$m, the spins can only flip near the border of the lattice, while the lattice remains antiferromagnetic in the bulk.

Theoretically speaking, the hydrodynamic theories on active turbulence remain to be sophisticated. One of the most challenging tasks is to implement reasonable boundary conditions that reproduce experimental observations. As we can see from our experimental results in comparison with the microfluidic experiment [13], the boundary conditions are crucial for understanding macroscopic behavior of active turbulence. Even slight differences in boundary conditions can change the types of the emergent order. In this sense, our experimental results can function as a touchstone for future theoretical works. As a matter of fact, our on-going study is focused on reproducing our experimental observation by combining a hydrodynamic theory and experimentally extracted boundary conditions, which should be published in the near future.

Our experimental results provide novel strategies for minimally invasive control of active matter systems to extract some kind of order out of the chaotic regime. This may be applicable to other experimental systems exhibiting vortex formation under geometrical confinement such as active colloids [19], cytoskeletal extracts [20, 21] and vibrated grains [22]. Self-organization of bacteria in nearly perfect vortex lattices can be used as a tool for more efficient energy extraction by an array of gears driven by swimming bacteria [23, 24], directed transport of objects [25], or control of turbulent active motion of bacteria in Newtonian [1, 3] or anisotropic fluids [26, 27].

5.5 Appendix

5.5.1 Properties of Reference Area

Here we summarize experimentally obtained properties of unconstrained bacterial turbulence in our reference area, the red rectangle region in Fig. 5.3.

We calculated the root mean square velocity v_{rms}, the mean vorticity, and the mean enstrophy defined as following.

$$v_{\mathrm{f}}^{\mathrm{rms}}(a) = \left\langle \sqrt{\langle [v_{\mathrm{f}}(r, t)]^2 \rangle_t} \right\rangle_{r \in \mathrm{ROI}_{\mathrm{ref}}}, \tag{5.13}$$

$$\text{mean vorticity} = \langle \langle \mathrm{rot}\, v(r, t) \rangle_t \rangle_{r \in \mathrm{ROI}_{\mathrm{ref}}}, \tag{5.14}$$

$$\text{mean enstrophy} = \langle \langle [\mathrm{rot}\, v(r, t)]^2 \rangle_t \rangle_{r \in \mathrm{ROI}_{\mathrm{ref}}}, \tag{5.15}$$

where $\mathrm{ROI}_{\mathrm{ref}}$ represents the ROI for the bulk unconstrained turbulence, which is shown as the red rectangle in Fig. 5.3. We also calculated the correlation function,

$$C_\infty(r) := \frac{\langle \langle v(r', t) \cdot v(r' + r, t) \rangle_{r' \in \mathrm{ROI}_{\mathrm{ref}}} \rangle_t}{\langle \langle |v(r', t)|^2 \rangle_{r' \in \mathrm{ROI}_{\mathrm{ref}}} \rangle_t}. \tag{5.16}$$

All the results are shown in Fig. 5.17.

Although the mean enstrophy gradually changes about ∼10% (Fig. 5.17c), v_{rms} stays almost constant, which assures that our experiment was done in a steady state (Fig. 5.17a). The mean vorticity shown in Fig. 5.17b naturally stays around 0, which means there is no a priori favored direction of rotations. This assures that the

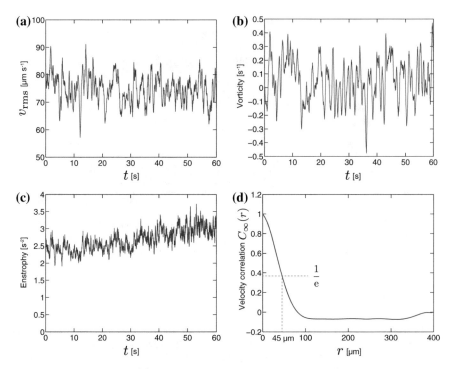

Fig. 5.17 Experimentally obtained properties of bulk unconstrained bacterial turbulence in the reference area ROI$_{\text{ref}}$ in the red rectangle shown in Fig. 5.3. **a** Time series of the root mean square velocity v_{rms}. v_{rms} is steady and we do not observe any discernible change. **b** Time series of the mean vorticity $\langle\langle \text{rot}\, \boldsymbol{v}(\boldsymbol{r}, t)\rangle_t\rangle_{\boldsymbol{r}\in\text{ROI}_{\text{ref}}}$. **c** Time series of the mean enstrophy $\langle\langle [\text{rot}\, \boldsymbol{v}(\boldsymbol{r}, t)]^2\rangle_t\rangle_{\boldsymbol{r}\in\text{ROI}_{\text{ref}}}$. **d** Velocity correlation function $C_\infty(r)$. Red dashed line represents the correlation length $L_\infty \simeq 45\,\mu\text{m}$ at which $C_\infty(r)$ becomes smaller than $1/\text{e}$

emergence of each vortex formed in the lattices is a consequence of spontaneous macroscopic chiral symmetry breaking.

From the correlation function $C_\infty(r)$, we can extract the correlation length of the unconstrained bacterial turbulence $L_\infty \simeq 45\,\mu\text{m}$. This value was compared with the correlation lengths in the lattice structures in Sect. 5.3.5.

5.5.2 Visualization of Vorticity

To gain insight into the bacterial dynamics, we visualized the vorticity field of bacterial velocity field as in Fig. 5.2a. Here we show an instantaneous vorticity field in Fig. 5.18b. For this purpose, we chose the range of the color plot as $[-5\,\text{s}^{-1}, +5\,\text{s}^{-1}]$ so that only 1.38% of the area in the field of view is saturated (Fig. 5.18a).

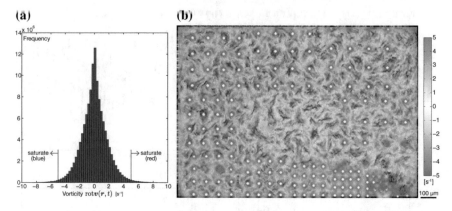

Fig. 5.18 a Histogram of vorticity calculated in all the field of view and in all the analyzed 2000 frames. We chose the threshold for the color map so that only 1.38% of the area is saturated. **b** Instantaneous vorticity field is overlaid on an experimental snapshot. Scale bar: 100 μm

References

1. Dombrowski C, Cisneros L, Chatkaew S, Goldstein RE, Kessler JO (2004) Self-concentration and large-scale coherence in bacterial dynamics. Phys Rev Lett 93(9):098103
2. Cisneros LH, Cortez R, Dombrowski C, Goldstein RE, Kessler JO (2007) Fluid dynamics of self-propelled microorganisms, from individuals to concentrated populations. Exper Fluids 43(5):737–753
3. Sokolov A, Aranson IS, Kessler JO, Goldstein RE (2007) Concentration dependence of the collective dynamics of swimming bacteria. Phys Rev Lett 98(15):158102
4. Sokolov A, Aranson IS (2009) Reduction of viscosity in suspension of swimming bacteria. Phys Rev Lett 103(14):148101
5. Sokolov A, Aranson IS (2012) Physical properties of collective motion in suspensions of bacteria. Phys Rev Lett 109(24):248109
6. Wensink HH, Dunkel J, Heidenreich S, Drescher K, Goldstein RE, Löwen H, Yeomans JM (2012) Meso-scale turbulence in living fluids. Proc Natl Acad Sci USA 109(36):14308–14313
7. Dunkel J, Heidenreich S, Drescher K, Wensink HH, Bär M, Goldstein RE (2013) Fluid dynamics of bacterial turbulence. Phys Rev Lett 110(22):228102
8. Wensink HH, Löwen H (2012) Emergent states in dense systems of active rods: from swarming to turbulence. J Phys Condens Matter Condens Matter 24(46):464130
9. Grossmann R, Romanczuk P, Bär M, Schimansky-Geier L (2014) Vortex arrays and active turbulence of self-propelled particles. Phys Rev Lett 113(25):258104
10. Dunkel J, Heidenreich S, Bär M, Goldstein RE (2013) Minimal continuum theories of structure formation in dense active fluids. New J Phys 15:045016
11. Wioland H, Woodhouse FG, Dunkel J, Kessler JO, Goldstein RE (2013) Confinement stabilizes a bacterial suspension into a spiral vortex. Phys Rev Lett 110(26):268102
12. Lushi E, Wioland H, Goldstein RE (2014) Fluid flows created by swimming bacteria drive self-organization in confined suspensions. Proc Natl Acad Sci USA 111(27):9733–9738
13. Wioland H, Woodhouse FG, Dunkel J, Goldstein RE (2016) Ferromagnetic and antiferromagnetic order in bacterial vortex lattices. Nat Phys 12:341–345
14. Wioland H, Lushi E, Goldstein RE (2016) Directed collective motion of bacteria under channel confinement. New J Phys 18(7):075002
15. Nishiguchi D, Aranson IS, Snezhko A, Sokolov A (2018) Engineering bacterial vortex lattice via direct laser lithography. Nat Commun 9(4486):1–8

16. Thiel M, Hermatschweiler M (2011) Three-dimensional laser lithography. Optik Photonik 4:36–39
17. Beppu K, Izri Z, Gohya J, Eto K, Ichikawa M, Maeda YT (2017) Geometry-driven collective ordering of bacterial vortices. Soft Matter 13(29):5038–5043
18. Hänggi P, Talkner P, Borkovec M (1990) Reaction-rate theory: fifty years after kramers. Rev Modern Phys 62(2):251–341
19. Bricard A, Caussin J-B, Das D, Savoie C, Chikkadi V, Shitara K, Chepizhko O, Peruani F, Saintillan D, Bartolo D (2015) Emergent vortices in populations of colloidal rollers. Nat Commun 6:7470
20. Sanchez T, Chen DTN, DeCamp SJ, Heymann M, Dogic Z (2012) Spontaneous motion in hierarchically assembled active matter. Nature 491:431
21. Wu KT, Hishamunda JB, Chen DTN, DeCamp SJ, Chang YW, Fernández-Nieves A, Fraden S, Dogic Z (2017) Transition from turbulent to coherent flows in confined three-dimensional active fluids. Science 355(6331):1284
22. Kumar N, Soni H, Ramaswamy S, Sood AK (2014) Flocking at a distance in active granular matter. Nat Commun 5:4688
23. Sokolov A, Apodaca MM, Grzybowski BA, Aranson IS (2010) Swimming bacteria power microscopic gears. Proc Natl Acad Sci USA 107(3):969–974
24. Thampi SP, Doostmohammadi A, Shendruk TN, Golestanian R, Yeomans JM (2016) Active micromachines: microfluidics powered by mesoscale turbulence. Sci Adv 2(7):e1501854
25. Kaiser A, Peshkov A, Sokolov A, ten Hagen B, Löwen H, Aranson IS (2014) Transport powered by bacterial turbulence. Phys Rev Lett 112(15):158101
26. Zhou S, Sokolov A, Lavrentovich OD, Aranson IS (2014) Living liquid crystals. Proc Natl Acad Sci USA 111(4):1265–1270
27. Genkin MM, Sokolov A, Lavrentovich OD, Aranson IS (2017) Topological defects in a living nematic ensnare swimming bacteria. Phys Rev X 7(1):011029

Chapter 6
General Conclusion and Outlook

Abstract General conclusion on the significance of our studies and the future outlook of active matter physics are given in this chapter. We summarize what the situation of active matter research used to be and how our experimental studies contributed to pushing forward the theoretical understandings of collective motion. Our results are discussed in the context of not only universality in the viewpoint of physics but also biological meanings. We conclude this thesis by suggesting possible future directions of active matter physics.

Keywords Collective motion · Active matter · Vicsek universality class
Active turbulence · Biological meanings

Throughout this dissertation, we have explored properties of emergent order and fluctuations in collective dynamics of swimming bacteria from the viewpoint of statistical physics. We have experimentally investigated the two major classes of collective motion: the Toner-Tu-Ramaswamy phases (the Vicsek universality class) and active turbulence.

Starting from the introduction of the Vicsek model in 1995, active matter physics has seen a great expansion of its community. As we have reviewed in Chap. 2, the large-scale numerical studies and corresponding hydrodynamic theories have revealed many fascinating universal properties of the Vicsek-style models, especially giant number fluctuations (GNF) in their homogeneous but highly fluctuating long-range ordered states with broken rotational symmetry (the Toner-Tu-Ramaswamy phases). However, those theoretical/numerical predictions have never been convincingly tested in experiments.

Inspired by these studies, 'GNF' were reported in many experimental studies. However, we pointed out that those 'GNF' observed experimentally were actually measured *out of the Toner-Tu-Ramaswamy phases*. All the reported 'GNF' were not deeply rooted in the mathematical properties of symmetry broken states, or the Nambu-Goldstone modes, but originated mainly from clustering or boundary effects. No GNF measurement had been done in long-range ordered phases before our study [1]. Due to theoretical and experimental pitfalls on 'GNF', there has been a widespread misunderstanding on GNF that 'GNF' have been trivially observed in experiments.

© Springer Nature Singapore Pte Ltd. 2020

D. Nishiguchi, *Order and Fluctuations in Collective Dynamics
of Swimming Bacteria*, Springer Theses, https://doi.org/10.1007/978-981-32-9998-6_6

Our experimental system on filamentous *Escherichia coli* bacteria has turned out to exhibit GNF in the true long-range ordered phase, as we described in Chap. 3. Therefore, it gives the first experimental realization of the Toner-Tu-Ramaswamy phases, and hence it falls into the Vicsek universality class. This is surprising because our bacteria with both complicated interactions and fluctuating internal degrees of freedom can even be reduced to the Vicsek universality class. Our finding of the first example of the Toner-Tu-Ramaswamy phenomena provides experimental grounds for many theoretical and numerical works.

Our results give many insights on the Vicsek universality class and on future directions of active matter physics. (i) By comparing our experimental system with other experimental systems and numerical simulations, we can extract what is necessary for the emergence of the Toner-Tu-Ramaswamy phases. The reasons why the Toner-Tu-Ramaswamy phases have been so difficult to observe experimentally or numerically except for the Vicsek-style models (e.g. simulations on rods with excluded volume) have remained elusive until our experiment came out. We might argue that the quasi-two-dimensionality (quasi-2D) associated with both adequate volume exclusion and reduced hydrodynamic instability is crucially important to make the system close to Vicsek-style. In this spirit, our experiment gives new ideas for future numerical and theoretical studies, such as testing quasi-2D simulations on self-propelled rods with excluded volume. As a matter of fact, very recent numerical [2] and experimental [3] works have supported this idea, and a new nonequilibrium mechanism for stabilizing orientational order in such a confined geometry has also been theoretically proposed [4]. (ii) Furthermore, our experimental results can contribute to the development of better theories. Only a part of our experimental results on the exponents of GNF and the correlation functions were consistent with the predictions by the Toner-Tu theory and the Vicsek-style simulations on self-propelled rods, which gives us many questions on the applicability of the theory and possible corrections required for application on real experimental systems. (iii) Because we have found the first system that belongs to the Vicsek universality class, we might be able to find conditions for other systems to fall into the Vicsek class by applying our knowledge that we have learned from our experiments. This is a very important future direction to understand the robustness of the Vicsek class.

So far, studies on collective motion, especially for the Vicsek universality class, were mostly numerical. However, because we have found the first example in that class, we can contribute and expect complementary development of theoretical, numerical, and experimental works.

In Chaps. 4 and 5, we reviewed and investigated the emergence of order out of chaotic bacterial turbulence of *Bacillus subtilis*. Although bulk *unconstrained* bacterial turbulence is well investigated, the interplay between bacterial turbulence and boundaries is far more difficult to treat theoretically, and we can still find unexpected phenomena of bacterial turbulence in contact with boundaries.

To address the question on what happens when bulk unconstrained bacterial turbulence encounters periodic structures, we devised an experimental system, using advanced microfabrication technique, in which we can observe bacterial turbulence in the structures in the different periodicity at the same time and for a long period of time [5]. As a result, we have succeeded to observe strong stabilization of vortices and emergence of the antiferromagnetic vortex lattice in a periodic pillar lattice with the lattice constant comparable to the correlation length of the bulk unconstrained bacterial turbulence. The emergence of the antiferromagnetic order was seemingly in contradiction with the existing study in microfluidic chambers [6], which highlights the importance of boundary conditions on macroscopic behavior of bacterial turbulence and its experimental investigation.

The study on boundary conditions in active matter is still in its infancy. Future studies should be focused on elucidating the realistic boundary conditions of active matter in general. This is of great importance if we think about applying concepts in active matter physics to much broader disciplines such as developmental biology and multi-cellular phenomena. During the development of organisms, cells collectively migrate according to some biological, chemical, or physical cues including boundary conditions and eventually form organs and the body, which then results in modification of boundary conditions. Such continuous feedback between the boundary condition and collective motion in active matter systems should be playing a significant role in life. Collective motion coupled with other degrees of freedom needs to be understood in the future.

As a possible future direction of active matter research, it is worthwhile to think about biological, ecological, or evolutionary meanings of collective motion. Interest of our studies so far was in extracting physics, especially universality, from collective motion that is distinctively different from individual behavior. Our studies as well as other studies on collective motion are sometimes far away from real biology. However, because we have provided some model biological experimental systems, future studies might fill the gap between active matter physics and biological science. Active matter physics still has a lot to learn from biology, and vice versa. We anticipate that both disciplines will coevolve into higher frameworks with broad applicability.

References

1. Nishiguchi D, Nagai KH, Chaté H, Sano M (2017) Long-range nematic order and anomalous fluctuations in suspensions of swimming filamentous bacteria. Phys Rev E 95(2):020601(R)
2. Shi X-q, Chaté H (2018) Self-propelled rods: linking alignment-dominated and repulsion-dominated active matter. arXiv: 1807.00294
3. Tanida S, Furuta K, Nishikawa K, Hiraiwa T, Kojima H, Oiwa K, Sano M (2018) Gliding filament system giving both orientational order and clusters in collective motion. arXiv: 1806.01049
4. Maitra A, Srivastava P, Marchetti MC, Lintuvuori JS, Ramaswamy S, Lenz M (2018) A nonequilibrium force can stabilize 2D active nematics. Proc Natl Acad Sci USA 115(27):6934–6939
5. Nishiguchi D, Aranson IS, Snezhko A, Sokolov A (2018) Engineering bacterial vortex lattice via direct laser lithography. Nature Commun 9(4486):1–8
6. Wioland H, Woodhouse FG, Dunkel J, Goldstein RE (2016) Ferromagnetic and antiferromagnetic order in bacterial vortex lattices. Nature Phys 12:341–345

Curriculum Vitae

Daiki NISHIGUCHI
Department of Physics, The University of Tokyo
Hongo 7-3-1, Bunkyo-ku, Tokyo, JAPAN.
E-mail: nishiguchi@noneq.phys.s.u-tokyo.ac.jp
Web: https://sites.google.com/site/daikinishiguchi/

POSITIONS:

Apr. 2019—present **Assistant Professor**
Takeuchi Laboratory,
Department of Physics, Graduate School of Science,
The Unviersity of Tokyo, Japan.

Jun. 2018–Apr. 2019 **Postdoctoral Researcher**
Group of Dr. Guillaume Duménil,
Pathogenesis of vascular infections unit,
Department of cell biology & infection, Pasteur Institute,
France.

Jun. 2017–Jun. 2018 **Postdoctoral Researcher**
Group of Dr. Hugues Chaté,
Service de Physique de l'Etat Condensé, CEA-Saclay,
France.

© Springer Nature Singapore Pte Ltd. 2020   127
D. Nishiguchi, *Order and Fluctuations in Collective Dynamics*
of Swimming Bacteria, Springer Theses, https://doi.org/10.1007/978-981-32-9998-6

Apr. 2017–May 2017 **Project Researcher**
                   Sano Laboratory,
                   Department of Physics, Graduate School of Science,
                   The University of Tokyo, Tokyo, Japan.

EDUCATION:

Apr. 2014–Mar. 2017 **Ph.D. in Science (Mar. 2017)**
                    Supervisor: Prof. Masaki Sano
                    Department of Physics, Graduate School of Science,
                    The University of Tokyo, Japan.
                    PhD Thesis:
                    "Order and fluctuations in collective dynamics of swimming
                    bacteria"
Apr. 2012–Mar. 2014 **M.S. in Science (Mar. 2014)**
                    Supervisor: Prof. Masaki Sano
                    Department of Physics, Graduate School of Science,
                    The University of Tokyo, Japan.
                    Master Thesis:
                    "Collective Motion of Self-Propelled Asymmetric Colloidal
                    Particles"
Apr. 2010–Mar. 2012 **B.S. in Science (Mar. 2012)**
                    Department of Physics, The University of Tokyo, Japan.
Apr. 2008–Mar. 2010 College of Arts and Sciences, The University of Tokyo,
                    Japan.

FELLOWSHIPS & HONORS:

Mar. 2020 **The 14th Young Scientist Award of the Physical Society
          of Japan**
Mar. 2017 **Research Award for Ph.D. students**
          Graduate School of Science, The Unviersity of Tokyo
Apr. 2014 – Mar. 2017 **Japan Society for Promotion of Science (JSPS)**
          Research Fellowship for Young Scientists (DC1)
Oct. 2012 – Mar. 2017 **Advanced Leading Graduate Course for Photon Science
          (ALPS) Fellowship, The University of Tokyo.**

Printed in the United States
By Bookmasters